Chaos and Determinism

The turbulent wakes of two parallel cylinders.
Photograph by R. Dumas, Institut de Mécanique
Statistique de la Turbulence, Marseille.

Chaos and Determinism

Turbulence as a Paradigm for Complex Systems Converging toward Final States

by Alexandre Favre

Henri Guitton

Jean Guitton

André Lichnerowicz

Etienne Wolff

MEMBERS OF THE INSTITUT DE FRANCE

Foreword by Julian C. R. Hunt, F.R.S., University of Cambridge

Translated by Bertram Eugene Schwarzbach

THE JOHNS HOPKINS UNIVERSITY PRESS
Baltimore and London

Originally © 1988 as
De la causalité à la finalité. A propos de la turbulence, in the
Collection Recherches Interdisciplinaires, series editor Pierre Delattre, by Editions Maloine, Paris, represented by The Cathy Miller Agency London.
Preparation of this English-language translation has been subsidized by the French Ministère de la Culture.

Printed in the United States of America on acid-free paper
04 03 02 01 00 99 98 97 96 95 5 4 3 2 1

The Johns Hopkins University Press
2715 North Charles Street
Baltimore, Maryland 21218-4319
The Johns Hopkins Press Ltd., London

ISBN 0-8018-4911-X ISBN 0-8018-4912-8 (pbk.)

Library of Congress Cataloging-in-Publication Data will be found at the end of this book.
A catalog record for this book is available from the British Library.

Frontispiece: Photograph by R. Dumas, Institut de Mécanique Statistique de la Turbulence, Marseille. Reproduced by permission.

Contents

Illustrations

Foreword
by Julian C. R. Hunt

When a somewhat lengthy calculation has conducted to us some simple and striking result, we are not satisfied until we have shown that we might have foreseen, if not the whole result, at least its most characteristic features. Why?

Because the lengthy calculation might not be able to be used again, while this is not true of the reasoning, often semiintuitive, which might have enabled us to foresee the result?

—Henri Poincaré

The great French physicist and applied mathematician Henri Poincaré (1908) was one of the first exponents of the need in science to understand and explain the general implications of scientific findings. He explained that this process is the first step in developing new lines of research. But it is also important for two other reasons: it enables the results of one scientific discipline to be understood by and perhaps used by those working in other scientific disciplines, and it is one of the essential steps in transforming research findings into practical applications.

Until recently, most researchers in mechanics thought there was no difficulty, in principle, in how they explain their results; they could describe the situation (e.g., satellites moving around a planet, fluid motion, etc.), what happens, which mechanisms are most important, and how they can be quantified and modeled. The Newtonian or Laplacian approach is usually thought to be sufficient for all situations, even where there are random phenomena.

However, this confidence of mechanics researchers has begun to be shaken (and also the confidence of outsiders who listen in on their arguments) by the realization that there are several quite different ways of investigating and explaining most complex problems in mechanics, whether it is interactions among five billiard balls, thick paint being stirred, or cracks in a random continuum. Should one attempt to describe in detail the causal or deterministic actions of each billiard ball or each eddy in turbulent motion (with particular attention to their initial conditions)? Or should one develop statistical models for an ensemble of essentially unpredictable interactions? Or should one look for global descriptions of the system's "state" as it evolves (e.g., the disorder of the system defined by its entropy or fractal dimension), or its sensitivity to disturbances, or its closeness to some "final" state (which in some situations may be when the maximum or minimum of some property occurs, such as energy, entropy, etc.)?

I write this Foreword on a mountain in California overlooking the San Andreas Fault. Although I cannot see or feel (at the moment) the random earth movements beneath me, I can see the random motions of the clouds as they form and dissipate in the wind from the Pacific: I can focus on their billows and their patterns of grouping, or I can watch in a general way the random filaments that are the final stages of a cloud as it evaporates. These observations correspond to two different methods of observing and analyzing random motions of fluids (or other systems) and, indeed, correspond to two different current approaches to research into complex unsteady fluid mechanics. These approaches differ in their methodology and to some extent in their philosophical basis. So it is not surprising that there is no consensus on which approach is valid or scientifically useful or on how to relate these different approaches.

Similar concerns about methodology and philosophy have arisen in other sciences and social sciences. So it was an inspired idea of Professor Favre, of the Institut de Mécanique Statistique de la Turbulence, in Marseille, to bring together five fellow members of the Institut de France to address these questions. As a result of Favre's leadership, all the authors familiarized themselves with the paradigm of turbulent fluid motion and used it as the basis for the discussion of other systems. The results of their deliberations over about four years are these eight essays, which focus on the dichotomy in analyzing complex chaotic phenomena between, on the one hand, causal explanations and, on the other hand, explanations in terms of the final "state" of a "system" (which by definition has many components, such as eddies, planets, etc.).

The title of the original version of the book, *De la causalité à la finalité*, gave rise to many misconceptions, especially among English speakers, about the meaning of the book. The French and English words *finalité* and *finality* each have two meanings, one of which is simply the final state. The alternative meaning of the French word—the aim or purpose of the final state—implies the existence of a teleological element in a system and is, therefore, somewhat metaphysical. The alternative meaning of the English word, *the principle of final cause in the universe*, has some of the same overtones as the alternative French meaning. To avoid any misapprehension that this book is not scientific, in this English version the authors use the word *teleonomy*, introduced by Jacques Monod to describe the relation between a state of a system and its final state or its convergence to a final state. This does not imply a metaphysical purpose, as the detailed studies in the book amply demonstrate.

Below is a brief summary of the key points in the eight chapters, along with some of my own comments and explanations.

Chapter I, Methods, Concepts, Vocabulary

Phenomena that are random in space and time occur widely in scientific systems and in many kinds of human activity. The key concepts for their analysis, many of which are mathematical, are described here in nonscientific language (as far as is possible). As is appropriate in a book that is both philosophical and scientific, theoretical definitions are given for *order* and *disorder*, and the methods used in their measurement are described, especially the statistical methods pioneered by Professor Favre's institute. Some of the most significant features of these phenomena have recently begun to be better understood by the study of unpredictable and chaotic dynamic systems (which may be, for example, as simple as pendulums oscillating with large amplitude).

Many natural phenomena and types of human behavior are chaotic. This is more than being simply unpredictable. For example, a person might walk unpredictably on one or the other side of the street, but if he or she wandered all over the road, traversing all possible paths, the movement would be chaotic. In mathematical language, the probability density function is finite, because there is an infinitesimal chance — unlike the first case, where there is a 50 percent chance — that the person is at any one point on one or the other sidewalk.

But a paradoxical aspect of many studies of chaotic phenomena is that there may be some order within them, for example, if our wanderer insisted on walking in random circular arcs (as we see in the chapters on fluid mechanics and biology). The recently developed techniques of fractal analysis have provided new methods for analyzing the order within chaos. A system can behave randomly within certain limits and with a certain pattern of behavior (within a basin of an attractor). Or it can quite suddenly change to another pattern (via a bifurcation). In our analogy, this would correspond to chaotic wandering on one or the other side of the street.

There is a particularly vivid description illustrating how dissipative systems are much more weakly correlated with their initial conditions than nondissipative systems. Consider a marble moving in a teacup; in the absence of friction, the marble oscillates forever, and in principle it would always be possible to calculate backward to find its previous trajectory if its present velocity and position are known. But if there is friction, the marble comes closer and closer to rest (according to the standard equation of mechanics), so that it becomes more and more difficult to measure its velocity and, therefore, more and more inaccurate to do the same backward calculation, although it is still possible in principle. This example shows why it is generally impossible (or highly impractical) to calculate changes of complex dissipative systems over significant periods of time either forward or backward using deterministic methods. It is generally possible only to estimate the statistics of such changes, using statistical assumptions or observations, the equations that govern the components of the systems (e.g., the motion of a single planet or an element in fluid), and any overall constraints on the system (such as energy or momentum conservation).

Approximate computational approaches have been developed for practical purposes based on heuristic, semiempirical models. However, Professor Favre emphasizes the more definitive contributions to our understanding, those that identify the most significant parameters and variables (by scientific insight or experience) and then use dimensional arguments to describe certain features of complex random physical systems. (For example, the height at which a turbulent chimney plume levels out in a stable atmosphere, such as in the early morning, can be estimated satisfactorily by such approaches, without any detailed understanding of the very complex processes in the eddies. These dimensional analyses, which do not give precise answers, are also essential for providing the framework for the effective use of model experiments to extrapolate to other conditions, usually on a large scale. Such experiments continue to be the basis for many engineering design calculations involving complex phenomena in mechanics.

Chapter I also briefly reviews some of the philosophical and logical concepts that arise in discussing chaotic complex systems, some of which can be defined in mathematical terms, like *causality*, *determinism*, *predictability*, and *state*, and some of which cannot be defined in mathematical terms, like *necessity*, *contingence*, and *freedom* (which means here the effects of human intervention on deterministic systems).

Chapter II, Turbulence in Fluid Mechanics

Many types of chaotic flow patterns and eddying motions occur at the onset of turbulence and in the fully developed states of turbulent flows. The authors emphasize the complexity of these flows, how they differ (e.g., between turbulence in pipes and turbulence in thermal convection), and also how mathematical equations can describe the flows. An example is given on the frontispiece of an experiment in a water flow. The illustration shows dye in the wakes of two parallel cylinders placed across the flow. As the large eddies of the two wakes interact, they destroy each other and form smaller eddies. However, in both large and small eddies there is visible order in the chaotic flow, a theme taken up later in the book.

Chapter II hints at the similarity of structure among different types of turbulent flow, a fact of great practical importance because it underlies most engineering calculations of turbulence: in small eddies, turbulent flows are similar because the eddies are in a state of local statistical and dynamic equilibrium, a concept first quantified by Andrei Kolmogorov (1941). Although the chaotic structure of large eddies has no general form, certain types of characteristic roller eddies are found in most turbulent shear flows (and would form downstream of those shown in the frontispiece); generally their axes slant upward or downward at about 30° to the flow direction (see Townsend, 1976).

Chapter III, The Atmosphere and the Hydrosphere

The atmosphere and hydrosphere form a stable interacting system over Earth's surface, whose state fluctuates only within narrow limits, a condition essential to the preservation of life. Turbulence is a particularly important feature of these environmental fluid motions, because it disperses and mixes heat and matter, thus performing a vital part of the purification of the atmosphere and the oceans. This global fluid system is essentially deterministic and can in principle be calculated, because its boundary conditions are known. The system is close to an optimum state, in the sense that the fluid motion of the present state, compared with other possible states, requires a minimum of energy to persist. (This is probably a controversial conclusion.) The authors do not discuss James Lovelock's (1979) alternative Gaia hypothesis—that the atmosphere and hydrosphere have evolved synergistically with the biosphere and that the latter is crucial to the maintenance of the long-term stability of the two other spheres. However, the authors certainly emphasize that the biosphere affects the atmosphere and hydrosphere and that freedom, that is, the intervention of the human, could possibly destroy enough of the biosphere to destabilize the equilibrium of the atmosphere and hydrosphere.

Chapter IV, Physical Theories

Two other important aspects of mathematical theory are introduced in chapter IV. Group theory, which is based on symmetry and invariance and which can provide a unifying explanation for some of the laws of physics, is also a powerful tool for analyzing possible final states of systems or for suggesting how equilibrium states can form. The calculus of variations is both an alternative analytical tool for calculating how a system evolves using causal analysis and a method for considering optimal states of systems.

Poincaré's point is here put another way: that any physical theory has two aspects, a *specialized form* couched in jargon and formulas and a *popular form* expressed in everyday language. The latter may be just as important as the former if the verbal analogies are valid, because it can stimulate the imagination of researchers. The popular form is also vital in educating and interesting the general public in science.

Relativity theory, statistical mechanics, and quantum mechanics are reviewed, using the concepts of group theory and calculus of variations, with emphasis on predictability, causality, and optimal states and on Heisenberg's question of how one chooses between the accuracy of measurement and interference in the object of measurement. A nice point is made that the *cause of a final state* of a complex system is a *convergence of individual causes.* In other words, the *final state* in a physical system is, in a broad sense, *deterministic.* The conclusion is that, for the study of physics (as I suggested when I was looking at clouds), there is no one correct approach; sometimes a search for causality is more fruitful than a search for a final state, and sometimes vice versa.

Chapter V, Biology

This brief chapter is concerned with *finalité*, or teleonomy, in the study of embryology. The question asked is, Are the laws of nature best examined causally by examining each phenomenon in terms of the whole system—or the final state of the whole system—even if the physics and the chemistry of such laws are not understood? Fascinating case studies are drawn from embryology and from the ways in which maimed animals can regrow limbs to precisely the same pattern as before, an excellent example of a teleonomic element in natural law. The second point of the chapter is to show that the existence of the self-replicating double-helix DNA module does not make biology a totally deterministic science.

Chapter VI, Economics, Turbulence, and Chaos

The behavior of complex systems is of great relevance to economic systems, the problem of bifurcation being of particular interest. In the study of economics (which still falls short of qualifying as a true science), it may be impossible to isolate elementary phenomena because of their many interlocking and complex aspects. Nevertheless, it is useful to make simple heuristic models. The analogy is drawn between turbulence and an economy undergoing disturbances yet still in statistical equilibrium, but caution is required in extending this analogy too far, because economic systems are very complex and their many interactions and semiautonomous elements can undergo qualitative and drastic changes or bifurcations. To consider economics as a science implies the existence of a natural order independent of human intervention. But then, what is meant by a *natural* order? Is it a state achieved only by deliberate action, in which case there is a contradiction? Or does it happen automatically? In economics as in biology there are advantages of teleonomic analysis in terms of some final state or system.

Chapter VII, Conclusion, and Chapter VIII, A Philosopher's Reflections

There are many different approaches to analyzing complex systems and even to analyzing different aspects of the same system, and they should be regarded as complementary. It does not appear that there is any general methodological or philosophical link or guide to how to connect the different approaches; the only current guide is to choose the approach (following chapter IV on physics) that gives rise to the most imaginative hypotheses and that enables hypotheses to be tested. These may or may not be the same approaches that give rise to the most useful application of the science.

For example, it has been found in turbulent flows that taking a causal or deterministic approach (in the sense of a nonstatistical and predictable approach over a short time) enables individual eddy structures to be analyzed and has given rise to new engineering designs for reducing the drag of boats or aircraft and for controlling mass transfer. For these purposes, this approach has proved to be more fruitful than those based on the statistical state of the turbulence, which may be either a statistical causal approach, based on equations that calculate changes in turbulence from a given state, or a teleonomic approach, based on the assumption that the turbulence is in local statistical equilibrium. But often the statistical analysis (e.g., using approximate equations for the mean movements of the velocity fluctuations) may be useful for analyzing the flow after the new device is installed. This is a practical example where the different approaches are quite complementary. (These points are discussed further by Julian Hunt and David J. Carruthers [1990].)

In general, it is desirable for scientists to consider the broad philosophical and methodological questions and procedures that guide scientific endeavor. Can it always be assumed that phenomena are intelligible? And which investigative concepts and methods are to be used? Is one looking for order or chaos? And is the explanation to be deterministic, or statistical, or in terms of the state of the system? What is the appropriate use of experimental controls? For researchers in traditional mechanics, determinism (defined in this book as the application of causality in science) is the main guiding concept.

Chaotic complex systems evolve in so many different ways, depending on their initial condition or small external disturbances, that even if they are evolving toward the final state there do not seem to be any reliable systematic ways of predicting this tendency or of knowing whether the system is in some final state, except in the empirical sense that it persists. The authors recount some of the interesting ways in which, from a "disordered" system, some ordered phenomenon can emerge (such as coherent structures in turbulence or chemical systems, as Ilia Prigogine [1980] finds). The only reliable predictive analyses of such phenomena are, it seems, causal, because there are few laws or principles governing how complex systems change and, in particular, how they move to a final state. (Notable exceptions are the second law of thermodynamics for isolated systems, kinetic theory of ideal gases, and conservative systems.) Although these laws or ideal examples do not apply to complex dissipative systems, they have stimulated researchers to construct approximate mathematical models describing the state of complex systems, on the assumption that they are close to some defined state (such as a statistical equilibrium) that persists even with external disturbances. This approach has been used in plasma physics and to describe two-phase flows (such as bubbles in turbulent liquid flows) and turbulent flows (e.g., Lumley, 1978). Such models are based on physical laws, but they always contain empirical elements. One could argue that the approximate equations for the moments of velocity components in fully developed turbulence are essentially equations appropriate for the local statistical equilibrium and are, therefore, another example of equations describing a final state.

This most unusual book has achieved its objective of breaking academic barriers and of showing that researchers of complex chaotic systems in different disciplines grapple with the same questions while using different conceptual and practical methodologies. They also share the same difficulties in understanding and explaining their results. The authors describe or at least mention almost every general idea current in mathematical and physical science. I found many aspects of my own thinking about fluid mechanics and the philosophy of science changing as I read the book and discussed it with colleagues and friends. I took away three main thoughts from my enjoyable month with this book.

First, those undertaking, assessing, and applying research into a complex system should consider carefully the deterministic features of the system and assess the unpredictable and possibly chaotic aspects of its behavior. This consideration should generally include both a causal (or a causal-statistical) analysis and an analysis of the system (which may also be statistical) in relation to some final state.

Second, the recent theoretical and experimental studies of complex systems in physical, biological, and social sciences have given us a new view of the world with philosophical, economic, and perhaps even political implications. From the eighteenth-century, Newtonian-Laplacian clockwork view of the universe created by some Providence who set the "circling planets singing on their way" (as the hymn writer expressed it), we moved during the first half of the twentieth century to a more statistical view, where Providence sets the boundary conditions, or rules, for the statistical systems, or dice. One might describe the midtwentieth-century view as the confident belief that the natural world is largely predictable and rational, so that with the assistance of information theory, computing power, and system control (based in part on the important concept of feedback), it would be possible even for the natural world to be controlled and utilized with ever greater sureness. At one point, hubris reached such a level that it was believed that weather, climate, and even oceans could be controlled by human intervention. This philosophy extended, in most countries, to belief in control of the economy through fine tuning and, in some countries, even to control of society. The prevailing view is now that complex systems of any sort are predictable only over short periods. They remain in or near stable states (or near attractors, which may or may not be optimum), provided they are not strongly disturbed. And if they are disturbed, large changes (bifurcations) can occur. These changes may have disastrous consequences for the global environment or great benefits for engineering problems. In late 1989, with the breakup of the Soviet Union, we saw how large changes can occur in a sociopolitical system as well.

Third, although complex systems pass through various states, perhaps in some cases to final states, there are no laws or even empirical algorithms for assessing whether a system (such as the world's climate or economy) is reaching a final state. So in science, especially in mechanics, the usual approach is still to apply causal or deterministic methods of analysis and experiment—but with the usual scientific aim of being as general as possible.

This book shows the merit of studying systems from the points of view of both causality and teleonomy; not surprisingly, one also learns that the scientific study of most systems remains biased more toward the former than the latter.

Preface to the French Edition

The moment seems to have arrived when researchers in hitherto independent scientific domains may begin to join forces to approach a common task. They are not accustomed to such cooperation. Some may even believe that the time is not yet ripe. Each domain has its own language, which is little understood by scientists outside the field. And in any event, much of the most productive research has come about as a result of the separation of the different fields of study. It is in his own laboratory or study that the scientist contributes to the progress of the science to which he has devoted himself. Why not continue this isolation in order to make further discoveries? It is, of course, not suggested here to do away with the separation of disciplines, which has proved not only fruitful but even indispensable. Our intention in this volume is to describe a complementary strategy.

After the reform of the French universities in the wake of the student revolt of 1968, there was much talk about interdisciplinary study and much difficulty in realizing it. It is an old dream. According to the archives of the Institut de France, Father Pierre-Claude-François Daunou proposed in 1795 that a single institute be created to shelter all intellectual activities. When the five academies of prerevolutionary France were brought together in a single organization, the Institut de France was formed, one of those rare institutions where the learned in many diverse fields might meet. In fact, however, except for the solemn annual opening ceremony, the academies of the Institut always meet separately and never deal jointly with broad issues of mutual concern. Modern life is so distracting and exhausting that it discourages the interdisciplinary contacts that the Institut de France was created to promote.

It was thus on a more modest scale that several members of sister academies of the Institut de France set out to work together, much as the learned men of the midseventeenth century had gathered in private meetings with Father Marin Mersenne, meetings that led in 1666 to the creation of the Academy of Science. In our case, we were simply friends, members of either the French Academy, the Academy of Science, or the Academy of Moral and Political Sciences. We were a small group and aware of our limitations, but we thought it necessary to begin somewhere, sometime, to initiate discussions, to ask questions, and to seek agreement on the basis of our varied scientific backgrounds and philosophical interests. We hope that our example will stimulate others to create similar groups.

The point of departure for our encounters was suggested by research on turbulence. It will become evident how significant the paradigm of turbulence proved to be for the study of complex systems. At first glance, it might seem that turbulence is an accident that disturbs some preestablished order, an accident whose effects ought to be attenuated and even eliminated in order to restore that order. However, the study of turbulence in general—and its application to the atmosphere and hydrosphere in particular—convinced us that turbulence is not an accident. What appears to be disorder is really not that at all, and in fact turbulence obeys laws of a deterministic character. This recognition led us to inquire whether there were not other domains in the study of inanimate matter—in physics, mechanics, or chemistry—where the same phenomenon might be observed. Better still, we asked ourselves whether living matter, the subject of biology and embryology, did not share the same tendency. Generalizing still further, we wondered whether social phenomena, in which so many variables interact and render observation difficult, might not also be subject to the same tendencies. It was necessary to ask whether analogies across these areas of inquiry are merely likenesses or, on the contrary, manifestations of a law that applies to everything that exists.

The idea of turbulence evokes ideas of complexity, causality, and teleonomy. These ideas are, in fact, intimately related. But the theories of bifurcations and qualitative dynamics were also essential for an understanding of the phenomenon of turbulence. This will be clear from the discussion throughout the book.

For more than four years we five met regularly and considered questions pertinent in each of our disciplines. We discovered that the questions of causality and teleonomy were in fact a common interest, but each of us had his own way of considering them. There was no consensus, and that was doubtless an advantage. The nuances that distinguished our approaches to these questions were due to the specific character of the areas and the nature of our research. We began to discover that our differences were not obstacles, nor were they contradictory. Quite the contrary, we discovered a clear convergence of our points of view, which became more and more compelling. This is precisely the advantage of interdisciplinary contact, which is rarely realized even within a single discipline.

We have written for two types of reader: those who are aware of these problems and those who are not but who want to know. We must therefore be not only audacious but modest and, especially, clear. We hope that all readers can follow the path that connects our separate chapters and that led to our conclusions. The chapters are not independent pieces strung together like a series of monologues. A physicist, a mathematician, a biologist, an economist, and a philosopher, we listened to our colleagues' presentations, extracting and expressing the principles that underlie the argument presented here.

While scientists and scholars continue to do their own research, it is essential to connect their questions, even the most up-to-date ones, with the general questions of philosophy. This volume is a first step. We hope it will encourage readers to further pursue this sort of inquiry.

Preface to the English Edition

This book is a philosophical reflection based on our research, each in his own scientific discipline and area. Chapters I through VII stay within the limits of scientific observation; chapter VIII offers a broader philosophical view regarding the issues raised in the preceding chapters.

The discussions and reviews that followed the publication of the French edition of this book have been valuable in translating it. In a meeting held at the Institut de France on 22 November 1988, we decided to try to emphasize in the English edition what may not have been entirely explicit in the French edition: that within the phenomenon of causality/teleonomy there is an epistemological distinction between the aspect that relates to science in the strict sense, and the aspect that is extrinsic to science, although still related to it. This book restricts itself to the aspect of causality/teleonomy that relates to science in the strict sense—to scientific observations and to the philosophical interpretations they suggest.

The concept of teleonomy in the French edition is invoked in Jacques Monod's sense (1970), as the nonmetaphysical part of teleology. The term was used before Monod in that sense by both C. S. Pittendrigh (1958) and Ernst Mayr (1961, 1974). Richard O'Grady and Daniel Brooks (1985) clarify the distinction between teleonomical and teleological activities. Teleonomical activities determine final states that "are reached because of internal controlling factors. [They are] 'end-directed.' Teleological activities have a purposeful behavior in which certain outcomes occur because events were deliberately brought about so as to produce them. This behavior requires some degree of cognition. It is most prevalent in humans and exists to various extents in other animals. The end state is a deliberately sought goal. . . . It is 'goal-seeking,' so as to make it clear that such behavior is the result of a consciousness capable of some amount of premeditation and choice."

We have introduced some other changes and, we hope, improvements in this edition. Illustrations have been added as well as additions to the text and the bibliography suggested by reviewers and readers, especially Paul Germain, Marcel Lesieur, Julian Hunt, Stephen Kline, and Marcel Barrère. Readers play a vital role in the scientific and philosophical dialectic, and we thank ours for the constructive and friendly spirit in which they have performed their valuable task.

Translator's Note

In principle, philosophical and scientific discourse is easily translated because philosophers and scientists are reputed to speak a precise and international language. In fact, however, there are national nuances and emphases, as a result of which even technical words can have different meanings.

Possibly because the scholastic tradition remained stronger and longer in Latin countries than in Anglo-Saxon countries that were influenced by the Reformation, France has retained scholastic terms for the four types of cause that Aristotle defined in the *Metaphysics* but did not name: *formal, material, efficient,* and *final.* While these terms are well known to English-speaking professional philosophers, their technical meanings are far different from the common associations these words suggest to nonprofessional readers.

In particular, *cause finale*, so obviously translatable as "final cause," is actually belied by that translation, which suggests some last cause in a chain of causes. In fact, a *cause finale* is not even a "cause" in the English sense of the word; its sense corresponds rather to the French and Old English *finalité*, finality (*fin*, "end," in the sense retained in the expression "means to an end"), which is best rendered as the objective, the purpose, for which a particular labor or project is undertaken.

Recently, an attempt to avoid both anthropopathic and metaphysical nuances has been made with the coinage *teleonomy*, from the Greek for "the law" (*nomos*) and "end" (*telos*), that is, a law for some end or objective. The biologist Jacques Monod (1970) used the word *téléonomie* to emphasize that *finalité* is scientific and nonmetaphysical. But the late Sir Peter Medawar (Medawar and Medawar, 1977, 11) objected to this word, pointing out that *teleonomy* is merely, from the etymological point of view, a "genteelism" for the religious term *teleology.*

Despite Medawar's objection, and to emphasize the authors' voluntary restriction of their argument to scientific models for these philosophical terms, I have, at the authors' insistence, generally translated *cause finale* and *finalisme* by one form or another of *teleonomy* in order to suggest the nonmetaphysical part of teleology that they are trying to define in a scientific context.

A similar problem arises for the translation of the scholastic *cause effi-*

cace, which has been retained in French but which is not a "cause" in the English sense of the word. Aristotle's "source of motion" might be best rendered as the immediate agent or as the agency by means of which changes are effected. But, at the authors' insistence, I have generally translated it as "efficient cause" in order to stay as close as possible to the French text.

B.E.S.

Chaos and Determinism

I

Methods, Concepts, Vocabulary

In interdisciplinary contacts, discussions reveal that differences in methods and concepts as well as ambiguities in language create obstacles to mutual understanding. These difficulties add to those intrinsic to the subject being studied. Therefore, even though we do not claim to deal with underlying epistemological and semantic considerations, we must set out conventions regarding methods, concepts, and language. We hope to be precise about the role, the characteristics, and the meanings of the terms we use.

I.1. The Role of Mathematics in the Natural Sciences

Mathematics is an abstract science that puts ontology between parentheses. It requires the mind to accept a sort of asceticism in order to purify reasoning and to expunge imprecision and inconsistency from discourse. The mathematical modeling of physical phenomena can thus ensure the latter a rational robustness for the analysis of their states and behavior. It is, however, clear that mathematical representation must be used cautiously to avoid overly simplified and reductive formulations, which risk being artificial and irrelevant. There are indeed many cases where ordinary language provides satisfactory description and allows valid prediction.

In the course of solving mathematical problems, one may succeed in finding rigorously exact solutions. But this is rather unusual and may have a purely formal character. It is more usual and often sufficient for most applications to obtain approximate solutions, where there is a trade-off between the closeness of the approximation and the time and effort that must be expended in calculations to improve upon it. One arrives at such solutions by means of series expansions and averaging techniques. With or without mathematical models, science never arrives at absolute truths—merely at truths approximated with greater or lesser degrees of precision.

This is recognized in the analysis of measure in quantum mechanics and statistical mechanics.

Mathematics essentially involves abstract analysis based on real situations; it is indifferent to the field of phenomena to which it is applied and treats them all in a single formulation. When a system is described by a mathematical model that is valid up to a certain level of approximation, this implies, and it may even be proved, that the system is completely determined (see I.5).

A description of a system that is more or less complete also provides the means of specifying the elements that can fully determine the states of the system. However, as we shall shortly see, even descriptions that are mathematically complete but local with regard to time may lead to fully determined behavior that is, nevertheless, in practice unpredictable in the long term. It may also happen that one is confronted by a mathematical model whose solutions cannot obviously be derived. Techniques of qualitative study have been developed that are capable of providing valuable information about the a priori qualitative behavior of phenomena. This may guide the experimenter by suggesting how to isolate these types of behavior, which would lead to finding the values of the parameters for which changes of state are observed.

When the equations are too complicated for solutions to be obtained by mathematical analysis or by numerical approximation, the measures of each term of each equation provide useful information regarding their relative importance. When certain terms cannot be measured, they may still be subject to estimation by their difference from the sum of the terms of a corresponding equation. This can suggest a choice of simplifying hypotheses.

However, the passage from a model consisting of theoretical equations to a model consisting of empirical equations is not without risk. Quite often we neglect the terms with numerical coefficients that experimental practice suggests are insignificant. A serious qualitative study can sometimes expose the risks of such approximations. Other simplifications are made when it would be too difficult to obtain solutions without them. In such cases, cruder models are constructed based upon reductive hypotheses suggested by intuition or experience. In principle, such hypotheses should be justified, a posteriori, by a comparison of predicted results with results confirmed by repeated experiments. Furthermore, it is necessary to check in each case that the hypotheses are both mutually compatible and compatible with the equations and properties to be studied, since simplified hypotheses may obstruct the observation of important properties. In the absence of a theory, empirical trials may be made to arrive at results that are applicable in analogous cases or, even better, in cases that satisfy laws of similarity (see I.9).

To study more complex phenomena, particularly in economics and biology, one can sometimes create mathematical models that are subject to verification by comparison of their results with repeated experiments. While in nature one cannot design experiments or observations to identify the role of a particular parameter, one can study the role of that parameter by simulation upon such a model.

I.2. Scales, Domains, and Methods of Observation; Structure, Stability, and Regulation

A phenomenon may appear quite different depending upon the scale of observation used. A fluid, for example, whether gas or liquid, observed by the unaided senses seems to be a continuous medium, while a fluid observed in more detail by suitable techniques and instruments shows itself to be composed of molecules and atoms and so is discontinuous. That the two representations of the fluid are not contradictory is explained by the different scales of observation.

For each study, it is necessary to determine the frame of reference, the scales, domains, and methods of observation, the properties that represent the phenomenon, and the parameters and the methods of analyzing the results of those observations.[1] When assuming the habitual three-dimensional representation of space, with time as a fourth, irreversible, dimension, *scale* is defined as the smallest volume within the interior of which one agrees not to try to distinguish the nonuniformity of a property being measured and as the shortest interval of time during which one agrees not to try to distinguish variations of a given property. The *domain* of observation is defined as the greatest volume and the longest time interval over which the study will be extended.

Many phenomena can be studied only by employing several scales of observation. The information gained from observations performed at one scale may be treated at a coarser scale, macroscopic in comparison with the first, by taking averages (by means of statistical techniques; see I.4) of larger volumes and longer intervals of time.

Since ancient times, atomist philosophers and scientists have sought ultimate reality in the structure of matter at increasingly finer scales in order to find "elementary particles," while astronomers have sought the structure of the universe in steadily widening domains. These studies have proven indispensable and fruitful, but those performed at intermediate scales and in domains directly accessible to researchers, such as studies of classic mechanics, physics, chemistry, and biology, have been no less useful, especially because they are convenient, more precise, and may be replicated.

Let us recall that, in everyday language, *structure* represents the way the constituent parts of a whole are arranged with respect to each other (Larousse, 1979). In philosophical language, on the other hand, *structure* means "a whole constituted of unitary phenomena in such a fashion that each one depends upon the others and may be what it is only . . . by virtue of its relation with them" (Lalande, 1983).[2] Relations depend upon time in evolving structures, and their conservation requires exchanges of energy within dissipative structures. It is in this sense that Ilia Prigogine (1982) discusses dissipative structures in physics, chemistry, and biology.

The study of turbulence in fluid flows (see chapter II) by means of statistical methods that analyze conditional spatiotemporal correlations (Favre, 1983; see also I.4) permits the discovery of turbulent structures disguised by very complex fluctuations. These evolving structures, which are at least partially dissipative because of the effects of viscous friction, are not always composed of the same materials. The methods mentioned permit a measurement of the celerity of the turbulent field, which is in general different from the average velocity of the fluid. (The equations for the spatiotemporal correlations show, in fact, that that difference is due to the average effects of nonlinear terms; see Favre et al. [1976]; Favre [1983].) The wave structure of a surge, for example, is propagated by its own celerity, even when the average velocity of the water is nearly zero. At the meteorological scale, the evolving vorticity structures of cyclones and large disturbances are partially dissipative and are propagated by their own celerity without always being composed of the same physical material.

For some mathematicians, structure is imagined as an axiomatic relation based on operations on sets, products of sets, sets of subsets, and the detailed properties for these subsets (Lichnerowicz et al., 1976). These properties correspond, for example, to ordering relations, composition laws, and topologies. The main example is given by the notion of *group* (a term that in mathematics has a very precise meaning). The notion of group is connected with the idea of invariance in the case of transformation groups (see chapter IV), where transformations do not change some quantities or properties. According to René Thom (ibid.), *structure* is defined as a spatiotemporal morphology described by significant spatial discontinuities and by the "syntax" that determines how these sets of discontinuities form into relatively stable systems.

In physics, as we shall see, a system is a set of interacting elements. The structure expresses the principles of that system's organization. Jacques Monod (1970) claims that modern biology is nearly completely structuralist. Having sought structural invariants first on an anatomical level and then at a cellular level, biology now studies invariance on the molecular level and the teleonomic behavior of living creatures in the structure of certain macromolecules (see chapter V). The structure of a living creature may be conserved, more or less, despite the renewal of its physical material.

In the domain of economics, the term *structure* has been used to refer to the proportions and relations that characterize an observable economic unit. Both evolve irreversibly in time (ibid.). Disturbing factors such as population pressure, inventions, innovations, and social mores act upon such economic structures as the hierarchies of a nation's industries and regions and disturb their functioning (for example, the balance of supply and demand). The inverse influences—of structures upon constituent elements—can be identified, although only after lapses of time and despite differing rhythms. It is necessary to specify the time scale and to distinguish analytically the parameters of function, structure, and social organization (ibid.).

According to André Lichnerowicz et al. (1976), it is necessary to distinguish, on the one hand, mathematical structures and, on the other, the structures that describe the fields of real phenomena. Many studies of real phenomena are characterized by

—a setting aside of certain properties in order to arrive at abstract structures, at grids;

—recourse to a global perspective, to a vast field of elements upon which a classification is introduced within which local phenomena are defined only in terms of global ones;

—the search for transformations with invariants for structures;

—the hope of detecting structures that are extensive in time and space, which poses the problem of structural stability; and

—the introduction of the concept of autoregulation, which perhaps implies another type of stability with respect to certain types of perturbations.

This leads to the idea of *stability*, a concept common to several disciplines and with many shades of meaning. In a stable system, small perturbations are spontaneously dampened. When, however, a system is unstable, perturbations create deviations that increase in amplitude and may attain very high values. The main practical distinction concerns material and form. The movement of a physical particle is stable if the control exercised by certain initial conditions permits the control of its position after a given lapse of time. This is *positional stability.*

It may also occur that an evolving system does not always act upon the same material particles but rather imposes a particular form upon different particles, a form that is conserved. This is *structural stability.* Such is the case for a surge. In the case of turbulent structures, structural stability is often apparent, and one can follow the paths of such structures over a considerable distance throughout their displacement and change, sometimes even after a wave of compression, or an expansion fan, or even a supersonic shock has passed through them.

In physics, a regulator is a mechanism that automatically guarantees the constancy of one or more variables during the functioning of the system, despite the perturbations experienced by factors that act upon those variables (Mathieu et al., 1983). In physiology, a regulator is a set of mechanisms that permits the maintenance of constancy for a function (Larousse, 1979).

According to René Thom (Lichnerowicz et al., 1976), one should define *regulation* as the set of mechanisms by which a system protects its existence along the time axis. Experimentation shows that, when systems are subjected to perturbations, they have the capacity of preserving their form, since spatial identity — or more precisely, space-time connectedness — is the basis of regulation. For the simple case of a ball dropped into a cup, the mechanism is stable, and one can hardly speak about regulation because there is no specific mechanism upon which regulation can depend. There would be such a mechanism if the cup were subject to deformation to the point where it acquired overhanging cliffs, which would correspond to an unstable state leading to the fall of the ball to the bottom of the cup, where it would find a stable position. In this case, regulation would be discontinuous. In embryology, the formation of a limb always begins by the acquisition of polarity, an abstract idea; only afterward does this limb acquire its bone and muscles, that is, its structure. Embryonic evolution thus passes from the abstract to the concrete (see chapter V).

I.3. Turbulence, Systems, Interactions, Nonlinearity, Attractors, and Bifurcations

It is natural to study first phenomena with simple properties and to follow these with progressively complex cases. The latter are composed of several diverse elements combined in a fashion not immediately obvious (Larousse, 1979), with oscillations that are either isolated or related and whose interactions render the entire phenomenon more than the sum of its parts (Lalande, 1983).

Complex phenomena are thus commonplace rather than rare; in physics, for example, turbulence in fluid flows is the rule and lack of turbulence the exception. A flow is turbulent when it includes a large number of eddies of varied dimensions. The frontispiece of this book shows the wakes of two

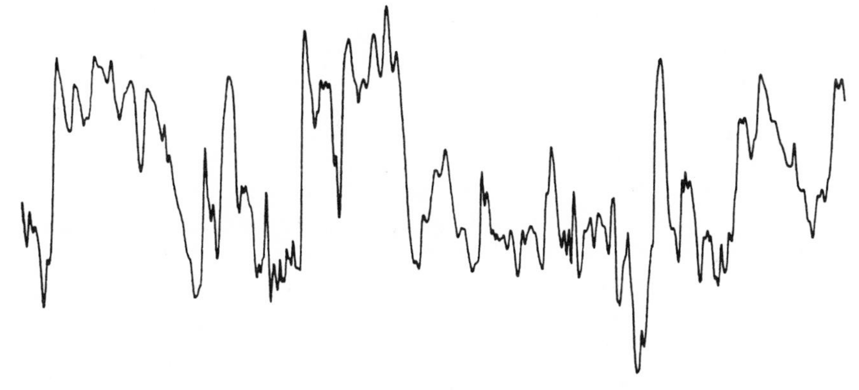

Figure 1. The record of turbulent velocity in a cylindrical pipe. From Favre et al., 1976.

cylinders orthogonal to a uniform flow of water, which develops eddies and turbulence. The flow of water, from left to right, has been rendered visible by the release of filaments of water colored by dyes, white upstream from the cylinders and red and green just downstream of each cylinder. Outside the zone of turbulence, the filaments diffuse slowly under the influence of molecular agitation. Within the wakes, they outline eddy structures and diffuse rapidly under the influence of the turbulence. The Reynolds parameter, defined as a function of the diameter of the cylinders, is $R_e = 2{,}000$.

A criterion of the complexity of a phenomenon is the amount of information necessary to define it and the practical difficulty in obtaining that information. For example, to define a linear phenomenon, two points suffice, for a circle, three points, and so forth. But to define a phenomenon as complex as turbulence, thousands of samplings are required to represent its properties during a single second at a single location within the domain, and it is desirable to increase both the number of locations in the domain and the interval of time to be observed. For instance, figure 1 represents the record of a sequence of longitudinal velocity fluctuations of the air flow in a pipe within a single location. The Reynolds parameter, defined as the function of diameter of the cylinder, is 35,000. The distance to the wall is 0.031 of the diameter, the mean velocity is 670 cm/sec.$^{-1}$, and the frequency of the sampling is 12 kHz. The complete measurements should also include the three components of velocity, fluctuations in pressure, and possibly temperature and density. The most difficult to measure are fluctuations in pressure.

The term *system* has acquired varied and sometimes contradictory meanings. We use the one frequently employed in thermodynamics (Mathieu et al., 1983): a system is taken as a set of material elements of a well-defined composition, localized in space and separated from the rest of the universe (called the *exterior medium*) by an ideal surface or a material interface. Depending on whether the interface is or is not impenetrable by material objects, the system is said to be closed or open. A closed system whose interface is also impermeable to heat is said to be *adiabatic.* If the system does not exchange material, heat, or any form of energy with the exterior medium, it is said to be *insulated.* At any given instant, the state of a system can be defined, with respect to the given scales, by the values of a number of variables—which may be great. If that state is capable of retaining its qualities indefinitely, even when it is insulated, it is said to be in *equilibrium.* A system that is not in equilibrium changes by undergoing transformations of various sorts, transformations classified according to the way they affect those changes. If a sequence of transformations returns the system to its original equilibrium state, that sequence is called a *cycle.*

A movement is reversible if the object that has moved can pass over the same path with the same velocity but with the sign reversed at each point. Reversibility is thus connected with invariance with respect to time reversal. Real transformations of macroscopic (with respect to human scales) physicochemical systems are naturally irreversible, always being accompanied by phenomena that lead to a dissipation of energy and that operate in a single temporal direction. They are associated with the creation of entropy, according to the second law of thermodynamics. Classic macroscopic thermodynamics employs the hypothesis of transformation by successive equilibrium states. Any dissipation of energy corresponding to the creation of entropy leads to a destruction of structures. But one also observes in nature the creation of animate and inanimate structures.[3] This is not incompatible with the second law of thermodynamics, because the production of entropy can continue even though the systems are not insulated, and the boundaries can be crossed by entropy fluxes. These phenomena may be explained by an extension of thermodynamics to the case where the transformation is produced far from the equilibrium state (see Glansdorff and Prigogine, 1971).

Regarding the evolution of biological systems, Henri Bergson (1954) observes that, the more deeply we analyze the nature of time, the better we understand that length of time implies invention, creation of forms, and continuous changes in what is new. The various parts of a system may interact not only among themselves but also with the system as a whole, which in turn affects its constituent parts. When the properties of a system are represented by equations with nonlinear terms, these terms express the significant interactions. *Nonlinearity* characterizes phenomena for which the mutually related properties do not vary proportionally. These phenomena are described by equations whose coefficients depend upon a function and its derivatives. Their solutions can no longer be superimposed.

The analysis of phenomena that change with time—whether in mechanics, physics, chemistry, biology, or economics—may be begun, at least qualitatively, by the methods of the theory of dynamic systems. To the extent to which these phenomena can be represented mathematically, we are led to a qualitative study of simple or partial differential equations, a domain to which Henri Poincaré contributed so magnificently a century ago. The application of these methods, particularly to the problems that arise from turbulent flows in fluids, has profited in recent years from developments inspired by the work of Edward Lorenz (1963), David Ruelle (1978), and Ruelle and Floris Takens (1971). Natural systems observed at our scale are most often dissipative—like mechanical systems with friction, which except for certain cases such as planetary motion are considered up to a first approximation. Throughout their evolution, almost all dynamic systems admit solutions that tend toward asymptotic limits, which are called *attractors.*

It is customary to present solutions in a phase-space whose axes represent coordinates of position and velocity or, rather, of moments linearly related to the time derivatives of position.[4] The changes in such a system are then represented by trajectories in phase-space. The conditions at the origin of those trajectories constitute a basin of attraction, if all the trajectories ultimately tend toward the attractor, whatever the initial conditions in their basin may have been. Complex systems often have several attractors, with several basins, and the phase-spaces may have more than two dimensions.[5]

Attractors come in many types, the simplest being the *point*, which corresponds to a solution independent of time, that is, a stationary state. Next is the *limit cycle*, which corresponds to a time-periodical solution characterized by an amplitude and a period. The *torus* describes a quasi-periodical situation in time that has several independent frequencies. Solutions to these equations may show singularities and possibly a loss of stability and a *bifurcation*, a term introduced by Henri Poincaré.

> Each time that, for a given value of a parameter, called a critical value, the solution of the equation or of the system of equations changes qualitatively, one says that there is a *bifurcation.* A point in the space of parameters where such an event occurs is, by definition, a bifurcation point. Two or more branches of solutions, whether stable or unstable, emerge from it. The representation of an arbitrary property, which is, however, characteristic of the single or several solutions as a function of the bifurcation parameter, constitutes a bifurcation diagram. (Bergé et al., 1984, 38)

Bifurcations assume diverse forms. We mention only the most common among them, the one-parameter bifurcation, which is normal (or supercriti-

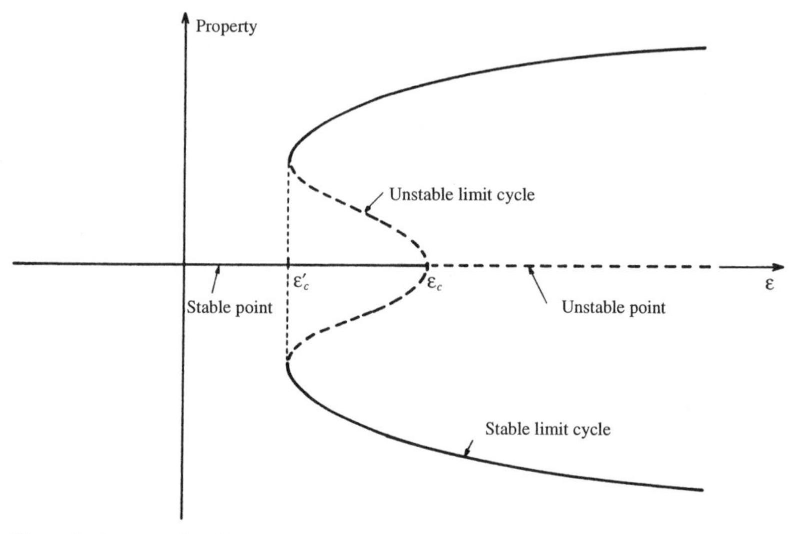

Figure 2. An example of inverse, or subcritical, bifurcation. From Bergé, Pomeau, and Vidal, 1984.

cal) or inverse (or subcritical), according to whether the lowest-order nonlinear terms of the equation reduce or amplify its instability.

The classic case of normal bifurcation occurs in the behavior of the undamped oscillator (the van der Pol equation). It dissipates energy while absorbing an average compensating energy from exterior sources. The trajectories in phase-space converge toward a stable limit cycle, which corresponds to a periodical solution. The bifurcation diagram represents the amplitude as a function of the physical parameter, ϵ. When this parameter attains the critical value, ϵ_c, the solution that corresponds to a stable point changes its property and is replaced at the bifurcation by a limit-cycle attractor with two symmetrical branches of stable solutions developed progressively from a zero amplitude and whose diagram has the shape of a fork.

Figure 2 is an example of inverse, or subcritical, bifurcation obtained by changing the sign of the nonlinear term in the equation just mentioned. When the parameter, ϵ, attains the critical value, ϵ_c, the solution corresponding to the stable point is once more replaced at the bifurcation by a limit cycle with two branches of symmetrical solutions, but the solutions are unstable from an amplitude of zero up to a limited amplitude of the considered property. These branches are tipped toward decreasing values of the parameter until a threshold value, ϵ'_c, is reached, when they tip in a contrary direction and thus achieve a stable cycle, which becomes the new attractor. These branches of periodically stable solutions are established only from situations with nonzero amplitudes.

Within the interval, $[\epsilon_c', \epsilon_c]$, there exist two stable solutions, but they cannot be satisfied simultaneously because they are determined by the initial conditions. Furthermore, when the parameter varies by increasing values, the transition between the singular point and the limit cycle occurs at the critical value, ϵ_c, whereas, in the opposite direction, it intervenes only at the threshold value, ϵ_c', because of a hysteresis phenomenon. These methods permit analyzing such complex aperiodic (or nonperiodic) phenomena as chaos (see I.7) and turbulence (see II.3) by examining, for chaos, types of attractor and, for turbulence, modes of transition between attractors.

To describe temporal chaos, David Ruelle (1978) and Ruelle and Floris Takens (1971) introduce the concept of "strange attractors," which describe systems highly "sensitive to their initial conditions." In such a system, by definition, two trajectories in phase-space that are initially arbitrarily close necessarily diverge from each other. These writers have shown that this may occur even for systems whose phase-space is small but at least equal to three. The concept of the dimension of a strange attractor may be introduced to measure the complexity of the phenomenon being studied.

An idea of the structure of certain strange attractors may, for example, be suggested by a rectangle stretched in the direction of its length and shrunk in its width, as a result of which its area is contracted to take dissipation into account. Then one folds it upon itself into a horseshoe shape. This operation is repeated a number of times to obtain the foliated structure of a strange attractor, and each iteration doubles the distance between points.

Among the modes of transition between attractors that lead to chaos, it is customary to mention the cascade of subharmonic bifurcations (see figure 3). When the parameter, μ, reaches a critical value, μ_1, the system undergoes its normal first bifurcation, the two branches of which constitute a cyclic periodic attractor. For increasing values of the parameter, a second critical point, μ_2, appears, where each branch of the first attractor bifurcates normally into two branches, and so two doubly periodic cyclical attractors are produced. This procedure is repeated to produce a cascade of bifurcations and attractors, whose number and period doubles each time. According to a general property,[6] this sequence converges so rapidly to an accumulation point that, after a small number of bifurcations (but greater than three), it can no longer be distinguished. Beyond this point, periodic attractors, corresponding to chaotic behavior, alternate.

Types of attractor and bifurcation, as well as modes of transition, are varied. In the case of turbulence (see II.3), certain transitions occur by cascades of bifurcations, while others occur directly. The shadowgraph in figure 4 shows a ¼-inch jet of carbon dioxide issuing into the air at a speed of 127 ft./sec. It is nonturbulent as it leaves the nozzle at a Reynolds number

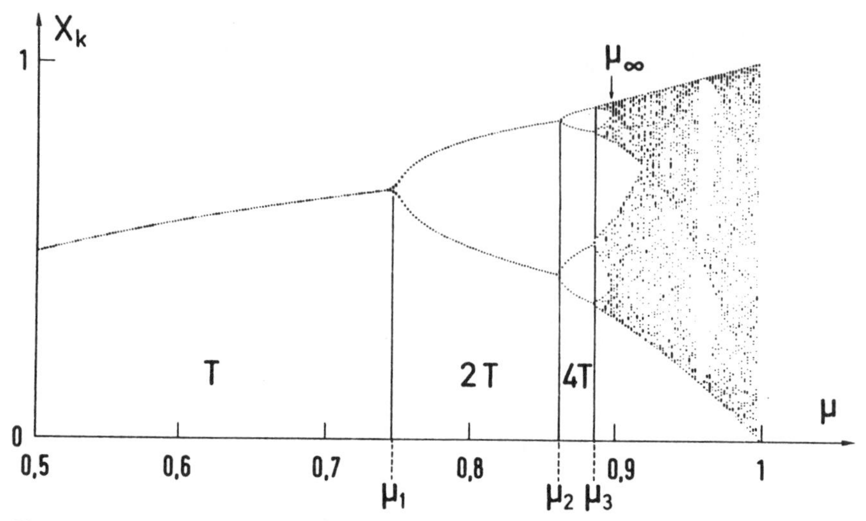

Figure 3. A cascade of subharmonic bifurcations. From Bergé, Pomeau, and Vidal, 1984.

of approximately 30,000. At 1 diameter downstream, it shows instability, the formation of vortex rings, and a transition to turbulence. These phenomena have not been entirely catalogued. Certain forms of transition may correspond to behavior still unknown in physics and biology, since the theory of dynamic systems is still developing.

Claude Bruter (1986) notes that there are two major difficulties in the study of such transition phenomena: the multiplicity of subobjects during destructuration and the extreme velocity of destructuration and restructuration, which masks the continuity of the transformation. Thus bifurcation is linked to the generation of distinct forms in a continuous universe. The ancient philosopher Empedocles anticipated the problem of transformations and bifurcations when he asserted that, at a given moment, "Entirety comes from Multiplicity, and at another moment it splits up, and the Entirety becomes Multiplicity."

Pierre Delattre (1986) uses the methods of the theory of dynamic systems to analyze the concepts of "efficient cause" and "teleonomy," which we discuss in a different context in this chapter (see I.5).

I.4. Statistical Methods, Averages, Fluctuations, Mean Effects, Correlations, Spectra, and Stages of Fluctuation

Very complex phenomena represented by a great amount of data often cannot, for practical reasons, be treated in all their wealth of detail. One then uses statistical methods that deal with ensemble averages or set averages of

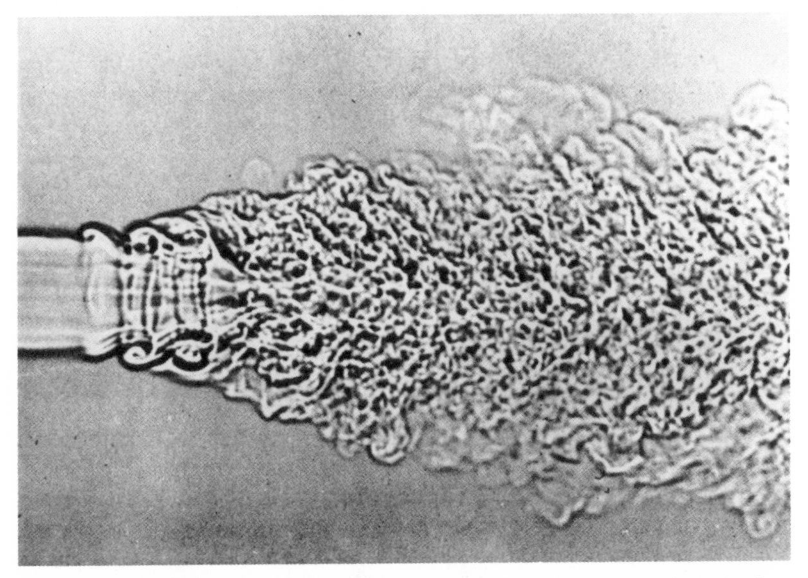

Figure 4. The instability of a round jet. Photograph by Fred Landis and Ascher Schapiro, from Van Dyke, 1982.

their various properties. The choice of statistical method does not in itself have philosophical bearing upon the concept of order or disorder of the phenomenon.

The differences between a measurable quantity and its average are called *fluctuations* when they are numerous and limited within the domain being considered. In nonlinear systems, the mean effects of fluctuations cannot, a priori, be neglected, because they may cause interactions capable of modifying the mean properties themselves. In effect, when one performs averaging upon the terms of equations, the nonlinear terms exhibit correlations that express these interactions.

The correlation of fluctuations is the average of their product. The corresponding coefficient of correlation is the correlation divided by the product of the standard deviations of the fluctuations.[7] According to the statistical method, one separates such fluctuating variables as A and B into their mean values, $\overline{A}$ and $\overline{B}$, and their fluctuations, A' and B', then one performs averaging upon the equations that relate them. The mean value of a linear term such as the sum, $\overline{A + B}$, equals the sum of the mean values, $\overline{A} + \overline{B}$. But the mean of a nonlinear term, such as the double product, $\overline{AB}$, is equal to the product of the means, $\overline{A}\,\overline{B}$, to which an averaging interaction term is added, the double correlation, $\overline{A'B'}$, which represents the mean effects of the fluctuations.[8] Finally, the mean of a derivative equals the derivative of the mean.

We have had occasion to show (Favre, 1946, 1965b, 1983; Favre et al.,

1976, pt. 3) that double spatiotemporal correlations are defined by considering two fluctuations at two points in space subject to a relative interval of time. This permits measuring statistically the mean celerity of a fluctuating field, following its propagation, and considering its statistical memory, the latter represented by the module of the maximal coefficient of spatiotemporal correlation as a function of the interval of time for a fixed spatial interval. The duration of statistical memory is the time represented by the area under the curve of statistical memory (see figure 6 and II.1.2).

In the special case where space separation is null, the correlation is exclusively temporal; and in the case where the interval of time is zero, it is exclusively spatial. If the intervals of space and time are both zero, then the correlation compares a fluctuation with itself under identical conditions and represents the mean of its square or its variance, and the coefficient of correlation equals 1. Correlation measures may be conditional when they select fluctuations according to their signs or thresholds; this permits the detection of random field structures, such as turbulence (Favre, 1983). A reciprocal correspondence is established by the Fourier transformation between the coefficients of correlation and the spectral densities, which exhibit the same information in different forms (Hinze, 1975).

In practice, the fluctuations can be decomposed by the classic Fourier method into elementary sinusoidal oscillations of either diverse frequencies with respect to time or diverse wave numbers in space. The contribution of the oscillations of each frequency or of each wave number can be represented as a temporal or spatial—or more generally, spatiotemporal—spectral function.

The spectra of periodic fluctuations include rays that correspond to frequency or to the basic wave number and their harmonics. The spectra of quasi-periodic fluctuations include rays at diverse, incommensurable frequencies or wave numbers. The spectra of nonperiodic fluctuations include oscillations whose frequencies or wave numbers belong to continuous intervals. In the case of complex fluctuating phenomena, the spectra may include rays, but they are mainly composed of wide bands.

The existence within one spectrum of zones separated by gaps, in which there are no oscillations contributing to the fluctuations, permits distinguishing a like number of *stages* of fluctuations, each of which is thus macroscopic with respect to the "lower" stages of higher frequencies and smaller wave numbers. Thus molecular motion within a fluid corresponds to a stage of fluctuation on a discontinuous medium, whose statistical means, on a human scale, represent fluid as a continuous medium. But we also perceive the motion of the vortices on the larger scale, which is characteristic of the turbulent fluctuations in the domain of fluid mechanics. On the still larger scale of the chain of meteorological observations performed by satellites within the planetary domain, one discovers the stage of large perturbations and the great currents of the circulation of the atmosphere and hydrosphere.

I.5. Causality, Determinism, Teleonomy, Necessity, Contingency, and Freedom

I.5.1. In its most common formulation, *determinism* is, as we have suggested, a property of a phenomenon whose current state is the consequence of its preceding states,[9] any of which determines its future states. Determinism also possesses a spatiotemporal character.

The concept of *causality* goes back to antiquity, where it was recognized by the Stoics. But Leibniz was one of the first of the modern philosophers to describe it in modern terms: "Nothing occurs without a cause or, at least, without a *determining* reason, which is to say, something that may provide an a priori reason why it should exist rather than not exist and why it should have this form rather than another" (Lalande, 1983, 127).

In a discussion of the reasons that lead to the appearance of an event, the occurrence of a phenomenon, Aristotle distinguishes four types of causes: material, efficient, formal, and final (*Metaphysics*), of which only the efficient and final causes have retained any currency in philosophical usage. *Efficient cause* refers to an agent, a phenomenon, that produces another phenomenon or, sometimes, to a being that produces an action, while *final cause* refers to the objective in view of which an action was undertaken (ibid.).[10] According to Kant, a final cause is an entirety that produces the existence of its own parts. Pierre Delattre (1986) discusses this question with greater precision in an argument suggested by the theory of dynamic systems and the correspondences with the concept of attractor basin. This can be summarized as follows:

Efficient cause expresses the relation of cause to effect for a local action affecting, step by step, in a temporal sequence, the constitutive elements of the system under consideration. *Final cause* is characterized by the system's global tendency toward some well-defined state (the *final state*), which may occur by whatever route, even when the conditions governing local causality at each point of the temporal sequence vary within certain limits. Subject to such a tendency, the change undergone by each part of the system depends upon those undergone by the other parts according to the reciprocal relations between local and global properties, which relations may include loopings and feedbacks. The relations between efficient causes and local properties, on the one hand, and final causes and global properties, on the other hand, lead to reciprocal relations between efficient and final causes. The representation of the change undergone by a system in phase-space that includes attractor basins illustrates this correspondence. Within the limits of an attractor basin, the initial conditions may be very different. They determine local causality and give rise to phase trajectories that are very different. But all of these trajectories converge globally toward a well-defined *attractor*, which corresponds to a *final state*.

Delattre further argues that, "when it is realized that efficient causality and final causes (teleonomy) can be closely correlated with local and global properties of the system studied, the apparent contradictions resolve themselves into complementary aspects in the description that can be given of the system. The concept of attractor basins as developed in qualitative dynamics is of prime importance in this respect" (Delattre, 1986, 154).

It is in biology that teleonomy appears most clearly. Even Jacques Monod (1970) concedes that it would be arbitrary and sterile to deny that the eye represents the accomplishment of a project, that of receiving images. All living beings are endowed with a "project," which he calls "teleonomy." The *Oxford English Dictionary* defines *teleonomy* as "the property, common to all living systems, of being organized towards the attainment of ends." We argue that the concept of teleonomy can also be applied to the behavior of inanimate material (see chapters II, III, IV, and VII), according to the principles of physics. We retain the concept of teleonomy as the nonmetaphysical part of teleology.

Determinism is the expression of causality in the domain of science (Bordas, 1983, 81). The different sciences have different ways of expressing determinism, and they recognize it in differing degrees, according to the language they use and the precision of their data. Mechanics and physics generally use the language of mathematics to formulate precise principles and laws. Mathematization is increasingly frequent in the other scientific disciplines, but it has not yet been accepted as a substitute for their specialized languages, especially as one passes from purely physicochemical phenomena to biological and human ones.

I.5.2. Let us recall the characteristics of determinism, as that term is applied to the physics of inanimate material (a definition we shall require again in chapters II, III, and IV). Jacques Hadamard expresses the fundamental principle of determinism in physics quite simply: "From the state of a (closed) physical system at the time, t_0, we may deduce its state at a later (but not at an earlier) instant, t" (Grünbaum, 1973, 234). According to Louis de Broglie (1937, 263), "For the mathematician, the determinism of natural phenomena is expressed by the fact that these phenomena are governed by equations whose solutions are entirely determined for all values of time, once one knows the values of certain quantities at a given, initial instant."

We now examine three forms of determinism: physicomathematical determinism; nonstatistical, experimental, physical determinism; and statistical, experimental, physical determinism. Throughout the rest of this study we reserve the term *physicomathematical determinism* for the cases in which both types of conditions that we are about to define are satisfied.

Physicomathematical determinism is characterized by two general conditions (first type) and three special conditions (second type) that describe the particular circumstances of the system, all five of which must be susceptible to precise formulation. The two general conditions of the first type are as follows:

1. The physical system under consideration is completely defined. It is composed of inanimate bodies whose mass distribution is defined and is bounded in space and time. Its state, observed at a certain scale in time and space, must be capable of being completely represented by the values of a finite number of quantities, the unknowns, which depend upon a finite number of independent variables, such as Cartesian coordinates in space and time.
2. The system changes by undergoing transformations governed by principles and laws that must be expressible by a number of equations at least as great as the number of unknowns, that is, a system of "closed" equations.[11]

The three additional special conditions of realization that signify the second type are as follows:

1. At any given moment, the system has boundary conditions at the limits (walls, free surfaces, closed reference surfaces) of its domain.
2. At any given instant, forces and events from the exterior medium (gravitation, electromagnetic forces, radiation, etc.) affect the system within the interior of its domain.
3. At any given instant, the state of the system, which is fixed by its initial conditions, may include a "memory" of preceding states.

Thus, when the two general conditions and the three special conditions for physicomathematical determinism are satisfied, which occurs when problems are well posed, the solutions of equations are completely determined, as are the states of the system that they represent, at all the subsequent instants and at every point in its spatiotemporal domain. Future states are determined for reversible systems as well as for irreversible ones. However, anterior states, although they may be determined for an ideal, reversible system, are determined with greater difficulty for an irreversible system with dissipation (Glansdorff and Prigogine, 1971; Grünbaum, 1973).

Consider a marble rolling in a cup. In the absence of friction, the marble will continue to move forever, and it will always be possible to calculate backward to find its previous trajectory if its present position and velocity are known. "But if there is friction, the marble comes closer and closer to rest, so it becomes more and more difficult to measure its velocity and, therefore, more and more inaccurate to do the same backward calculation, though still possible in principle" (Hunt, 1990, 387). But once the marble has come to rest, it becomes impossible to calculate its previous trajectory.

I.5.3. Physical determinism may also have a purely experimental origin, in which case the following conditions must be satisfied, because this type of determinism is based upon the observation that a phenomenon is repeated each time the same experiment is performed. The general conditions are (1) the physical system must be completely defined, as in the preceding case, and (2) the system may undergo change by means of transformations even though their laws may not be known and need not be mathematizable.

The special conditions for this type of determinism are the same as those for physicomathematical determinism, but in addition, the experiments or observations must also be susceptible of replication. Thus, if the state of the system replicates itself in the course of each (identical) experiment or during each observation, at each subsequent instant and at each point within the domain, the system may be considered experimentally deterministic up to the limit of the precision of measurement. The parameters that govern the behavior of the system belong to a domain that guarantees the structural stability of the system.

For complex and fluctuating physical systems, it is difficult to design and perform experiments that can reproduce the initial conditions exactly, but it is sometimes possible to choose a simple nonfluctuating initial state, even if the system eventually becomes complex and fluctuating. The consequences of the preceding states, the system's "memory," will then be replicated.

I.5.4. Experimental physical determinism may also have a statistical character, which is based upon the replicability of a phenomenon, up to a statistical average, for a series of experiments or observations. The conditions for this type of determinism are the same as those for experimental determinism, if we qualify the definitions of the state, the properties, the boundary, and the external and initial conditions of the system by the statistical averages that correspond to a large number of experiments or observations for each series of experiments. The series should be replicable under conditions that are close to identical.

If the states of a system are replicated for each series of experiments, at every subsequent instant, and at every point in the domain, the system may be considered to be statistically determined up to the precision of measurement.[12] Turbulent fluid flows are subject to this type of determinism, as are molecular motion and numerous phenomena of high-energy microphysics (Grünbaum, 1973). Statistical determinism obviously does not prove that a system is subject to strict determinism, although conversely, a strictly determined system would exhibit statistical determinism. This is the case for turbulent fluid flows.

I.5.5. Consideration of determinism leads to an examination of its relationship with necessity, contingency, and freedom, which are very complex concepts, whose usual definitions we shall repeat here and then translate into the more precise, scientifically oriented language we are developing.

In the *Metaphysics*, Aristotle defines *necessity* as follows: fundamentally, what is necessary cannot be otherwise; a particular consequence is inevitable once the principle is conceded (Lalande, 1983, 676).

Contingency is, in all respects, the contrary of necessity: all that is conceived of as capable of either existence or nonexistence, in any respect, or subject to any qualification, is said to be contingent (ibid.).

Freedom, as opposed to necessity, is "the possibility of effective choice among several actions according to their nature and/or consequences. . . . It is formally compatible with even rigorous determinism with respect to scientific knowledge" (Bordas, 1983, 159).[13]

If we compare these definitions to the preceding definition of physicomathematical determinism, which is the most rigorous form of determinism, it is clear that (1) the general conditions stated above correspond to *necessity*, and (2) the special conditions of a given situation correspond to *contingency*. Thus defined, physicomathematical determinism is compatible with both necessity and contingency, as they are usually defined, as well as with the *freedom* that may be exercised within the limits of contingency.

In practice, the general structural conditions described in the preceding section first require a complete definition of the system within the scale being considered. Only after that very stringent conditon is satisfied may the quality of necessity be claimed. Those definitions further require that the system be governed by principles and laws of a general nature, which cannot be contravened and are, thus, necessary. Such is the case for physics and its general principles of conservation of matter, energy, and momentum as well as the laws governing the behavior of the particular materials being studied.

For a fluid, for example, general equations represent all its possible states, while its behavior necessarily obeys the principles and laws of physics.

Since the principles of physics are variational, they are formulated in terms of partial derivative equations. Their solutions, or integrals, involve arbitrary, and thus contingent, constants and functions that correspond to the special conditions pertaining to a particular system. These conditions are a mathematical statement of constraints of the boundary conditions of the system, such as (for a fluid system) walls, free surfaces, and interfaces. These conditions are contingent and have an important effect on the solution; for example, they depend on whether the interfaces are permeable or impermeable to matter and energy. Air may flow, for example, within a cylindrical duct. But it may also flow around and within the surface of an object such as an airplane, whose surfaces include such deformable parts as the rudder and ailerons. The latter lend the air flow "contingency" and, if the rudders and ailerons are controlled by a pilot, even "freedom."

Special conditions may also include the influences of contingent forces whose origins are outside the system. Thus electromagnetic forces, gravity, and radiation may or may not impinge upon the system and may or may not vary with time. Gravity may exercise an influence when the system approaches a star and may disappear when the system has traveled a sufficient distance from that star. To pursue our aeronautical example, the commands that control the rudder of a plane may be transmitted from afar by radio in a contingent or even arbitrary manner. The state of a system at a given moment, its initial conditions, may also be contingent. At the moment of takeoff, the position of the rudder, the speed, and the power of the motor may all differ according to the prevailing particular conditions, and they may be freely chosen by the pilot.

Let us consider a wider domain than that of physics. If someone wants to travel from his residence to his workplace, and if he has several equally possible itineraries, there is contingency. An itinerary will not be chosen at random but will always be determined when the person makes a choice among them. The reasons for that choice may in fact be hidden and even unconscious, such as those determined by his previous history, or they may be conscious, such as those motivated by considerations of optimization. The determinism in such a situation is compatible with some general purpose, represented by an objective — arrival at the workplace — the accomplishment of which is the reason for this act, by whatever itinerary it may be pursued.

I.5.6. Determinism also poses the question of predictability. When a phenomenon is determined, its future states are fixed and ought, in principle, to be predictable. So it is in simple cases. For example, exact solutions have been obtained by elementary mathematical analysis for the equations describing the free fall of an object in a vacuum.

But there are complex phenomena that are mathematically represented by equations whose solutions are so complicated that, in practice, they cannot be extracted precisely by the available techniques of mathematical analysis or numerical calculation, even using the most powerful computers available. In addition, errors are inevitable from the outset of calculations because of the imprecision of the information concerning the initial state of the system, approximations introduced by simplifying hypotheses — the rounding off systematically performed to the smallest scales of the spectra of fluctuations to reduce the number of calculations. These errors are compounded in the course of mathematical operations and contaminate the results increasingly as the interval of prediction lengthens. Thus for certain circumstances there is unpredictability beyond a certain point in time.

For these reasons, the atmosphere and hydrosphere constitute a very complex system; it lends itself to prediction of future states with a precision that decreases rapidly when the interval between observations and the state to be predicted grows from several hours to several days (Lorenz, 1969a, 1969b). The fact that a complex phenomenon cannot be predicted exactly in the long term with the techniques now available is not incompatible with its being fully determined by the principles of physics.

I.5.7. In order to represent the behavior of a matter at the scale of human observation, classic mechanics has principles and laws, which are expressed by systems of closed equations. The very complicated phenomena studied here, including turbulent flows in fluids (chapter III), are still subject to physicomathematical determinism. "The mechanical sciences are in general deterministic; if a system under consideration is perfectly defined, if the initial conditions, positions, and velocities are precisely known, if the forces acting upon the system are known (as functions of time, position, and velocities), its movement is uniquely determined" (Académie des Sciences, 1980, 191).

Even phenomena whose velocity is not negligible with regard to that of light remain subject to determinism, as explained in Einstein's theory of relativity. With regard to causality, an event may act upon another only by signals and forces propagated by a velocity equal at most to that of light. This is geometrically represented by the interior of a cone; no actions from the exterior of the cone are possible (see chapter IV).

The deterministic principles of classic mechanics remain valid as well for the study of phenomena on the molecular scale. In 1950 the discipline of molecular dynamics began to develop thanks to powerful computers (Hansen and Lévesque, 1985) which made it possible to integrate numerically the coupled equations of interacting molecules and then to take averages to determine their macroscopic properties. But in general, one still employs the methods of statistical mechanics, which permit determining the statistical functions that describe macroscopic properties.

In microphysics, Heisenberg's uncertainty principle is thought to have led to a questioning of determinism but not to a denial of causality—which in modern physics is understood to be the assertion that all interactions must produce some effect in the future. It is in this sense that quantum mechanics remains causalist (chapter IV).

Quantum mechanics has redefined the concept of physical states; they are no longer described by a single point in space and a single moment of time but rather "in a more abstract manner, in terms of mathematical functions satisfying certain requirements . . . and [has] thereby restored the causal character of physical analysis" (Margeneau, 1978, 49).

I.5.8. Let us now discuss Laplacian determinism (Lalande, 1983; Laplace, 1921; Margeneau, 1978, 41).

> We must, therefore, consider the present state of the universe as the effect of its preceding state and as the cause of the one that will follow it. . . . An intelligence knowing at a given instant of time all forces acting in nature, as well as the momentary positions of all things of which the universe consists, would be able to comprehend the motions of the largest bodies of the world and those of the smallest atoms in a single formula, provided it were sufficiently powerful to subject all data to analysis; to it, nothing would be uncertain; both future and past would be present before its eyes. The human mind offers, in the perfection that it has been capable of lending astronomy, a weak sketch of that intelligence. . . .
>
> Man's discoveries in mechanics and geometry, together with those of universal gravitation, have made it possible to understand, using the same analytical expressions, the past and future states of the world. Applying the same method to several other objects within the realm of his recognition, he has come to attribute observed phenomena to general laws and to foresee to what the given circumstances will give rise. All these efforts in the search for truth tend to bring it increasingly closer to the intelligence, which we have just imagined, but it will always remain infinitely short of that mark. It is this tendency that is so special to mankind, that renders men superior to animals, and it is progress of this sort that distinguishes nations and centuries and that constitutes their glory. (Laplace, 1921, 3–4)

We cite this passage, which René Thom (1980, 122) quotes in extenso from Laplace, to show to what extravagant philosophical interpretations concerning, in particular, human action and liberty this extreme expression of absolute and universal determinism may lead, although Laplace himself did not draw such conclusions from them. André Lalande's analysis suggests some of them: determinism is "the philosophical doctrine according to which all the elements of the universe, and in particular human actions, are connected in such a fashion that, things being what they are at a given moment in time, there is only one state that, for that time, is compatible with the previous or future state" (1983, 221–22). This philosophical doctrine is, in effect, based upon the universal and absolute physical determinism that Laplace considered at the level of an intelligence that will always be infinitely higher than human intelligence. But the question of knowing whether the behavior of the universe satisfies the conditions for physicomathematical determinism (see I.5) *cannot be settled,* because a knowledge of the entire body of pertinent information is lacking.

One must first try to define the universe. According to the *Dictionnaire de Physique* (Mathieu et al., 1983, 535),

> It is the set of all existing bodies: planets, stars, nebulae, galaxies, clusters of galaxies, and atomic or subatomic particles of intergalactic matter. Currently, the exploration of the universe extends to distances of about 20 billion light-years. . . . The universe swims in isotropic, electromagnetic radiation that corresponds to that of a black body at 3°K (3 degrees above absolute zero, or −270°C). The most satisfactory model of the universe describes it by means of Riemannian space-time, whose geometric properties are related by Einstein's general relativity equations to the distribution of matter-energy.

But we can neither define the universe completely nor describe the set of all its constituent bodies with their composition, nor do we know whether it is delimited in space and time. We know neither whether its state may be represented by a finite number of properties depending upon a finite number of independent variables nor whether the principles and laws that govern it may be expressed by as many equations as, or more equations than, the number of unknowns. Are there conditions imposed at every moment at the limits of its domain? And do remote influences exerted by the exterior medium exist? And are they significant? What are its initial conditions?

How can living beings be described? As René Thom (1980, 123) observes, "There exist in macroscopic reality at our scale enormous blocks of phenomena whose verbal description is satisfactory but for which a mathematically rigorous description of the Laplacian type would be not only very difficult but not even pertinent. Such is the case in particular for the description of living creatures." It would not be enough to know the fine structure of a telephone exchange, which is infinitely less complicated than that of the human brain, to know the content of the messages passed through it.

Laplacian determinism, understood as an extension of physicomathematical determinism from the limited domains where it has been successfully asserted to the entire universe as well as to human actions, is *not valid,* because such an extension is not warranted by any proof and cannot, in any event, be applied on such a scale for lack of the requisite complete information about the universe and the human psyche; thus the question of whether Laplacian determinism can accommodate both contingency and free choice is not pertinent. While physicomathematical determinism is meaningful when applied to a limited and well-defined system, where it still leaves a margin for contingency and choice, it is far from obvious that under Laplacian determinism there could be scope for either, since it is not likely there will be reruns of a "parallel" universe established by different initial conditions—which, in any event, are not expressions of human choice, free or constrained.

1.6. Order and Disorder

"Order is an idea fundamental to intelligence. . . . In its most general sense, order is an expression of temporal, spatial, and numerical determinations, series, correspondences, laws, causes, objectives, genera and species, social organization, moral, juridical, and aesthetic norms, etc." (Lalande, 1983, 720). In a discussion of this definition, D. Parodi says that, "in all the above definitions, there is at least one idea in common, that of an intelligible relation. By virtue of that, order becomes the contrary of disorder, where we recognize only a state of fact without being able to extract from it any sort of definite relationship. This may be only an illusion, and every state in disorder may indicate only our confused knowledge of it, or our absence of knowledge."

This idea of disorder is subjective.[14] Following Dominique Parodi (ibid., 720–21),

> The relation discovered among the objects of thought may be more or less intelligible, whence degrees of order, and it may be intelligible in several different ways:
>
> First, in the sense that the place of an element is exactly determined with respect to the others, even if we do not see a clear reason why its place should be there rather than elsewhere. For example, the order of the numbers in the expression of π.
>
> Second, in the sense that the place of each element appears to be determined by a general reason, in conformity with some principle of causality or some law.
>
> Third, in the sense that the place that each element occupies is not only subject

> to determination or intelligible with respect to a particular relation but that this relation itself seems to be rational, satisfies wit and sentiment, including some reason for its being quite particular to itself and, most often, some purposive value, whence the ideas of social, moral, and aesthetic order . . . and finally, the metaphysical ideas of absolute order.

As we have already seen, when there exists a closed system of equations, with initial conditions, boundary conditions, and forces originating in the exterior medium, which correspond to a state of being, the phenomenon those equations describe is subject to determinism and obeys relations of a complete order that has an objective quality.

When the properties of a system may be expressed in the language of probability, we hold the condition for complete disorder to be the independence of those properties, according to which the probability of any one event is unchanged by the occurrence or nonoccurrence of all the other elements comprising the system. In other words, for all pairs of properties, A and B, the conditional probability of A if B occurs is in fact equal to the simple probability of A,[15] and this is true for diverse values of the parameters, particularly for spatiotemporal ones. The condition for complete disorder is thus, in fact, objective.

When the properties of a system may be represented by fluctuating variables, a sufficient condition for the system not to be disordered is that the correlations, and in particular the spatiotemporal correlations, do not vanish.[16] Thus, regarding the numerous complex and fluctuating phenomena in nature for which physicomathematical or experimental determinism has not been established—there would have to be such phenomena to lead us to entertain the hypothesis of disorder—in order to exclude the hypothesis of intrinsic disorder, it suffices to discover conditional probabilities different from the simple probabilities—or even nonvanishing correlations. In the absence of such information, the system may be considered apparently disordered, with the qualification that this appearance may possibly be due to the inadequacy of our knowledge of it.[17]

Scientific and philosophical discussions of this subject still continue. But even if a phenomenon is characterized by complex fluctuations, that does not imply, a priori, that it is subject to disorder.

I.7. Disordered Chaos and Deterministic Chaos

I.7.1. The word *chaos* has acquired various meanings according to the period and the discipline in which it has been used, ranging from the popular meaning of disorder through molecular chaos to deterministic chaos. Since it is often used in scientific and philosophical discourse, it seems appropriate for the purposes of this work to render a precise meaning and even to reconcile its diverse meanings.

"Chaos, coming from the Greek, χάος, chasm or abyss, was, according to Plato (Banquet), the dark, unbounded void that preceded the actual world, but it does not seem to have had the character of an eternal reality. According to Genesis, it was a confused mixture of all the elements of the world before they were structured by an organizing power. Subsequently, the word was understood to represent a disordered and heterogeneous set" (ibid., 138).

I.7.2. Since the nineteenth century, we have used the term *molecular chaos* in physics to designate the "state of complete disorder of positions and velocities that prevails in a mass of a perfect gas in thermodynamic equilibrium" (Mathieu et al., 1983, 64).

However, molecular motion does obey the deterministic principles and laws of classic mechanics. The fact that this phenomenon is treated by statistical methods because of its great complexity does not imply that it is disordered. These methods in fact permit calculation of macroscopic properties at the human scale, such as mass per unit volume, pressure, and temperature.

In the more general case where molecular motion is in neither mechanical nor thermodynamic equilibrium, these methods also permit calculation of friction and thermal conduction, which both have dissipative effects.[18] The irreversibility of these exchanges of momentum and thermal energy, which is characterized by statistical entropy, is statistically justified by the fact that reversibility is highly improbable when the molecules are numerous (Dodé, 1979, 29).[19]

I.7.3. In fact, in certain areas of physics and mechanics, complex fluctuating phenomena are described as being in *deterministic chaos* (Bergé et al., 1984) when it can be proved that they satisfy the conditions of determinism (see I.5), while their behavior still shows the characteristics of chaos, which are outlined in the following section (I.7.4). In fact, the term *chaotic* is used in geophysics, chemistry (see Libby and Williams, 1980), and other disciplines to describe complex, fluctuating phenomena that are nevertheless representable by systems of deterministic, closed equations. The qualitative theory of nonlinear differential equations shows, in effect, that such systems can have complex, fluctuating solutions sufficient to be considered chaotic while remaining strictly deterministic (Swinney and Gollub, 1981; Bergé et al., 1984).[20]

Edward Lorenz (1963) gives an example of a deterministic, nonperiodic flow—spectra with wide bands (see I.4 for definitions)—with three nonlinear, ordinary, differential equations representing a simplified model of a fluid flow, due to thermal convection, between parallel horizontal planes. The solutions obtained by numerical calculation may in fact turn out to be chaotic (see II.3 and figure 9).

I.7.4. To define chaos, David Ruelle and Floris Takens (Ruelle, 1978; Ruelle and Takens, 1971; see I.3) introduce the idea of sensitivity to initial conditions, where, in a phase-space, two trajectories that are initially arbitrarily close necessarily diverge from each other. The attractors are strange and have fractal dimensions and foliated structures, in which case the phase-space must have a dimension of at least three. Chaotic phenomena are nonperiodic, and their temporal spectra must be mainly conposed of one or several wide bands that represent the contribution of oscillations belonging to continuous intervals. The temporal correlations, which are the Fourier transformations of these spectra (see I.4), tend toward zero when the interval of time tends toward infinity.

To reconcile the various meanings given to disordered and deterministic chaos and, in particular, to generalize their use to describe systems whose properties fluctuate not only with respect to time but also to position in space, we propose that chaos is experimentally defined by the fact that the spatiotemporal correlations of the fluctuations, taken in pairs with time intervals and spatial displacements, tend toward zero when the time interval tends toward infinity and the spatial displacements tend either toward infinity or toward the maximum dimensions of the domain (see figure 6 and II.1.2) (Favre, 1992, 324). In *disordered chaos*, the correlations vanish for all intervals of time or spatial displacements that are themselves not zero.[21]

In the situation where the correlations tend toward zero when the interval of time and the spatial displacements tend toward infinity but do not all vanish for finite intervals of time and spatial displacements, the chaos is not disordered. We have seen (I.5) that deterministic chaos is that chaotic situation that can be represented by a closed system of equations with initial, boundary, and external conditions—or when it is replicable. The Fourier transformation of these spatiotemporal correlations then corresponds to the spectra, which are principally composed of wide bands in time and space. They represent nonperiodic phenomena.

To simplify the exposition, we may consider only the statistical memory of the fluctuating phenomenon defined above (I.4) as the coefficient of the maximal spatiotemporal correlation for each fixed spatial displacement. Disordered chaos is characterized by a null duration of the statistical memory, while the fluctuating phenomenon that has a statistical memory whose duration is not null but is limited would be chaotic but not disordered. It is deterministic if it can be represented by a closed system of equations with boundary, initial, and external conditions, or if it can be replicated. The above criteria are consistent with the general use of either disordered or ordered and deterministic spatiotemporal chaos, while eliminating the periodic and quasi-periodic phenomena whose correlations do not tend toward zero for infinite displacements in time and space and that are thus not chaotic.

The transition from a nonchaotic situation to a chaotic one may occur in several fashions. A typical scheme is that of a cascade of subharmonic bifurcations. It is composed of normal bifurcations with doublings of the period. Beyond three bifurcations, they accumulate in such a fashion that it becomes difficult to distinguish them. A zone of considerable complexity is formed where there is a mixing of periodic and nonperiodic attractors: subharmonic and inverse cascades in which the period is halved. A chaotic state is soon attained (see figure 3).

Turbulence in the flow of fluids (see II.3) is composed of fluctuations that are not only temporal but also spatial. They correspond to spatiotemporal chaos. The transition from nonturbulence to turbulence may be produced by either various mechanisms with bifurcations or directly and intermittently.

I.8. Chance and Randomness

In common language, *chance* (in French, *hasard*, from the Arabic, *az-zahr*, a game of dice) is "the fictitious cause of events that appear to be exclusively subject to the laws of probability. Unforeseen events, good or ill fortune" (Larousse, 1979, 499). This sense is distinct from that of disorder, which obeys no law at all. *Randomness* (in French, *aléatoire*, from the Latin, *alea*, a throw of the dice) "depends upon an uncertain, chancy event. It is a variable that can assume any number of values, each of which is associated with a certain probability" (ibid., 28). Aristotle defines *chance* as an "accidental cause of exceptional or accidental effects that bear the appearance of finality" (Lalande, 1983, 402).

According to Henri Poincaré, *chance* is the quality of a rigorously determined event of a type where a minute difference in its causes will produce a considerable difference in the final effect (ibid., 407). For example, an increase of one part in a thousand in the impulsion lent to a roulette ball will make it come to rest upon a different number.[22] In such a case, the law of large numbers should be considered as a derived property that is the result of these two conditions and of the postulate that the probabilities of the two causes themselves vary according to a continuous function. Henri Bergson (1907, 254–55) explains, "the particular options of the mind when it seeks to define chance. . . . It hesitates," he says, "incapable of decision, between the idea of an absence of causal agency and the absence of purpose."

We follow Antoine Cournot, for whom *chance* is the quality of an event "induced by the combination or coincidence of phenomena that belong to series that are independent with regard to causality" (Lalande, 1983, 405). Cournot does not discuss purpose in this context. In any event, it is possible that there may be still unrecognized connections among these causal series. Using Cournot's definition, the probability of an event induced by a coincidence of several phenomena belonging to independent series is equal to the product of the probabilities of each series. This product tends toward zero rapidly when the number of series increases, and the more numerous the series, the more strongly their coincidence shows that they are not independent, so the occurrence of events by reason of a coincidence of numerous phenomena cannot be explained by chance.

According to François Jacob (1976, 329–30), "that evolution [of mammals] should be due exclusively to a succession of microevents, to mutations each of which arises by chance — time and arithmetic argue against it. To extract from a roulette wheel, throw after throw, subunit by subunit, each of the several hundred thousand protein chains of which the body of a mammal is composed, would require a length of time that far exceeds the age of the solar system." René Thom (1980, 120) writes that "asserting that chance exists is to claim that there are natural phenomena that we shall never be able to describe and thus never be able to to understand [and] amounts to staking out an ontological position. . . . For the scientist, it is a duty, to be refused only at the risk of self-contradiction, to adopt an optimistic stance and to assume that nothing within nature is, a priori, unknowable."

In scientific discussions, the word *random* is used in preference to *chance*, and even more so in English, where *chance* is a rather ambiguous term. "In each of a series of experiments that are as identical as possible, a *random* quantity may take a value that is basically unknown but still determined by the result of the experiment. Depending upon whether the set of possible values is denumerable, the random quantity may have a distribution that is discontinuous (discrete) or continuous as a function of a parameter that may, in fact, be time. A random or stochastic process is a random function of time" (Mathieu et al., 1983, 10).

Random phenomena are usually studied by statistical experimental methods that use probability calculus to analyze the results. One considers a set composed of all the objects that manifest a certain form of a certain trait, for example, that $N = 1{,}000$ balls contained in an urn. These objects may be distinguishable by some mark, such as a number. In this set, a property, X, of these objects is called random if it can take diverse forms, x_i, which should be denumerable, $x_1, \ldots x_i, \ldots x_k$, when the distribution is not continuous, or nondenumerable when the distribution is continuous. The probability, p_i, of the discrete form, x_i, is the ratio of the number of objects having that form to the number of objects, N, in the set; this ratio is a precise number.[23] For example, if X represents the weight of the balls, each of which may be 100, 200, or 300 grams, and if 400 of the balls in the urn weigh 200 grams each, the probability of drawing a 200-gram ball would be 0.40.

If a complete inventory of the set cannot be performed in order to calculate the properties defined by the probabilities, one uses the statistical method that consists of experimentally removing a series of N' objects from the set in such a manner that each sampling should be independent of the others — or at least that it should not modify the probability laws. One then measures the frequency, p_i, with which the trait, x_i, appears, which yields the ratio of the number of objects with that trait with respect to the number, N', of objects in the series. The frequency is a variable that may take diverse values in different series, so it is random. For example, if one removes 100 balls from the urn, replacing each one and then mixing all the balls before removing the next, one may find that the number of 200-gram balls is unpredictably different from 40 in each of the series; it is therefore a random number.

The law of large numbers, discovered experimentally and proven in the calculus of probabilities, expresses the fact that frequency, p_i', tends toward probability, p_i, when the number, N', of experiments in a series tends toward infinity.[24] Thus random phenomena and chance obey laws of a statistically deterministic or global character.

The proof of the law of large numbers (Bass, 1967, 93–106) is based upon three conditions. The first is that the set to be studied be probabilizable and completely defined, which implies in our example, that 1,000 balls are indeed contained in the urn. The second is that the samplings from that set be mutually independent. The third is that a sufficiently large number, tending toward infinity, of samplings be taken. Since the number, N, of objects is limited, it is nearly certain that each object withdrawn and replaced in the course of a sampling will in fact be withdrawn many times. This amounts to admitting that the set is inventoried only partially, and in disorder, but over so many times that the average result is nearly what it would be if the set were completely inventoried a single time. This is an intuitive explanation of the fact that (since both the set and probability are determined) frequency measured on the basis of a large number of experiments is statistically determined. Experiments confirm this hypothesis.

When the terms *chance* and *randomness* are used in everyday language they do not lead to many ambiguities and misunderstandings; but in philosophical and scientific discourse, the word *chance* is ambiguous, and it is preferable to avoid it unless it is sufficiently qualified to indicate a precise meaning. In the sciences, the term *random* is frequently used to qualify complex fluctuating phenomena. Statistical measurements and probability calculus are frequently used to study these phenomena, but the choice of these methods does not imply, a priori, a philosophical position relative to the deterministic or disordered quality of the phenomena.[25]

I.9. Analogies and Similarities

When objects or systems manifest one or several qualities in common, one says that there are, in order of increasing force, *likenesses*, *analogies*, *similarities*, and *identities*. In everyday language, however, *analogy* is merely a stronger or weaker *likeness*. The original meaning of *analogy* is the identity of a relationship that identifies pairs of traits of two or several pairs of objects. In particular, and especially, the mathematical proportion that was called *analogy* in Euclid is now called *similarity* (Lalande, 1983, 53, 992).

According to Aristotle (*Rhetoric*), *metaphor* is the transfer of a name that designates one object to a different one: transfer of a designation of a genus to a species, or of a species to a genus, or from species to species, or according to some relation of analogy. By analogy, he understood that the second term was to the first as the fourth was to the third, because poets would use the fourth for the second and the second for the fourth and because sometimes one would add that term to which the word replaced in the metaphor refers. For example, there is the same relation between *old age* and *life* as between the *evening* and *day.* A poet could speak of the evening, as did Empedocles, as "the old age of the day" and of old age as "the evening of life." There is an unbroken scale between literary embellishments, like "the light of the mind," and profound ontological analogies, like Leibniz's "human virtues are analogous to divine perfections."

From the scientific point of view, it is preferable to use the more precise and restricted term *analogy.* We say that there is an analogy between systems that may be of different categories if they are represented by closed equations having the same formulation and if their initial conditions, boundary conditions, and the forces that act upon them from outside the system are formulated in the same way. The behavior of phenomena of a certain category may then explain, at least qualitatively, that of other categories. When all of these conditions are not satisfied, the analogy is only partial. For example, an analogy exists between the distribution of the streamlines of a fluid assumed to be frictionless as it passes about a given object and the distribution of the lines of electric current in a conductor as it passes about a nonconducting object of the same form. In fluid flows, there also exist partial analogies, called Reynolds analogies, between heat and mass transfer and momentum (the product of volumic mass and its velocity; see Favre et al., 1976, 141, 149, 171, 343).

The geometric similarity of two figures when they differ from each other only in their scale of length — that is, when the one can be derived from the other by homotopy (similar triangles, etc.) — has been recognized since antiquity. More generally, *mathematical similarity* is defined for a category of systems represented by the same closed equations, with the same initial conditions and exterior forces acting upon the system, and with boundaries of the same form (geometrically similar interfaces). With this in mind, we first express the equations and conditions in dimensionless terms, which are thus independent of the units of measurement. Lengths, times, and masses will then be related to the lengths, times, and reference masses characteristic of the phenomena in question. For example, to study the airflow around geometrically similar airplanes, one relates the lengths to the cord of the wing, the duration of the flight of the plane over that length (or its velocity), and its mass to that of ambient fluid. The equations would then have the same terms without dimensions and with nondimensional coefficients, called similarity parameters, which may take different numerical values for the different systems, according to their own properties.

When the similarity parameters are equal, the equations of the systems are identical and their solutions are the same in nondimensional terms. This condition is sufficient for these systems to be mathematically similar.[26] In mechanics they would be dynamically similar. There is, in such cases, an identity of relation that unites such homologous terms as *forces* and *trajectories.* The solution for one case, be it theoretical or experimental, is valid for all cases similar to it.

Within a single category of deterministic physical systems (subject to the same exterior forces, derived from similar initial conditions, and exhibiting the same forms at their interfaces), it is not necessary to express the equations to find the parameters of similarity. They may be obtained by dimensional analysis (Monod-Herzen, 1976). In fact, all physical laws should be expressible in a form independent of the units of measurement employed. The method consists of choosing quantities that govern the category of the system under consideration and of determining their combinations without dimensions, which is what constitutes similarity parameters.

In the example of geometrically similar airplanes, the governing quantities are the forces of inertia and the viscosity of the fluid, whose ratio constitutes the Reynolds parameter of similarity (Hinze, 1975; II.3).[27] Experiments with models in wind tunnels and on prototypes in flight thus permit predicting the behavior of all airplanes for which the given conditions of similarity are satisfied. Thus, when computational techniques are inadequate to obtain complete solutions for the equations, it may be possible to develop theoretical models, but these generally require experimental verification. However, in some cases, no such model is practical, and then empirical trials have to be performed. Similarity considerations permit predicting the behavior of phenomena belonging to a given category on the basis of the known behavior of another member of the category. In such a case, it is sufficient to describe the phenomenon quantitatively as a function of similarity parameters.

Notes

1. Unless otherwise specified, for the frame of reference we assume the axes associated with Earth and take into account the weight and the Coriolis inertial force. A domain may, under certain circumstances, subsume other domains that do not belong to it.

2. Following Claude Levi-Strauss, one can say that a system's structures are relations, such as correlations, that render the organization of a system intelligible, expressing its abstract properties.

3. These creations and transformations of structures by fluctuations are deterministic procedures in the case of turbulent fluid flows (see chapter 2).

4. One can speak of all the possible states of a system as a space of states; this is quite different from the use of the word *phase* in physics when it refers to the solid, liquid, or gaseous forms (or states) of the substance.

5. A phase-space has a dimension double the number of degrees of freedom necessary for a complete description of the system.

6. The difference between successive critical values of a parameter is reduced each time by an almost constant factor, 4.559 . . . , and the scale of distances is reduced by a different constant factor, 2.502. . . .

7. The standard deviation is the square root of the mean squares of the fluctuations.

8. The fluctuating variables, A and B, are thus decomposed (Hinze, 1975): $A = \overline{A} + A'$, and $B = \overline{B} + B'$; thus $\overline{A'} = 0$ and $\overline{B'} = 0$. The mean of the double product exhibits a double correlation of the fluctuations:

$$\overline{AB} = \overline{(\overline{A} + A')(\overline{B} + B')} = \overline{AB} + \overline{A}\,\overline{B'} + \overline{A'}\,\overline{B} + \overline{A'B'} = \overline{A}\,\overline{B} + \overline{A'B'}.$$

The mean of a triple product exhibits double and triple correlations.

9. This is what introduces the concept of memory.

10. The term *cause* may also refer to a chain of causes in irreversible time (Grünbaum, 1973, 181–83).

11. A differential system, which is composed of equations relating a variable *t*, a certain number of functions and their derivatives up to order, *n*, can be reduced to the case where only first-order derivatives appear by increasing the number of unknown functions and equations. A differential equation of order *n* is equivalent to *n* first-order, ordinary differential equations. As for first-order partial differential equations, their integration is analogous to that of first-order ordinary differential equations. In certain cases, one can express the general integral of second-order, linear, partial differential equations by means of arbitrary functions. In any event, to integrate second-order, partial, differential equations one must have recourse to initial and limiting conditions. These conditions remain valid when the order of the equations is higher than two in the analytic context.

12. According to Bernouilli's rules, the experimental average in a series of experiments converges as fast as the square root of the number of experiments, N', when it is large, and the average of the experimental averages obtained in the course of performing *S* series of N' measurements converges *S* times faster. It is thus proper to perform a large number, N', of experiments in each series, but it is sufficient to perform a moderate number of series, *S*, of experiments.

13. In physics (Mathieu et al., 1983, 116–17), one defines the number of degrees of freedom according to "the number of coordinates that fix the position of a mechanical system and that can vary independently, taking interrelations into account. A physical point has three, a rigid solid has six (three translations, three rotations)." For a continuously deformable medium, it is "the number of variables necessary for a complete determination of the state of the system." This number may be extremely high for a turbulent flow, as high as 10^{18} for the atmosphere.

14. Bergson (1907) cites the example of his library, into which someone has introduced an order, although he had been able to find documents more easily by memory before that order was installed.

15. This is true except for a phenomenon compared with itself under parametrically identical conditions, for which the probability of *A* if *A* occurs is obviously equal to one. The same is true for the correlation coefficient of *A* with *A*.

16. This condition is not necessary, because, within ordered systems, there may exist some null correlations (Bass, 1967, 63). For completely disordered systems, they are, on the contrary, all null.

17. All elements might obey the same probability law and yet be independent. For example, in the case where that law is Gaussian and centered and where the correlations are null, there is independence.

18. The system is out of equilibrium when gradients of average velocity and temperature exist.

19. Statistical entropy is proportional to the logarithm of the number of states that might appear in the molecular arrangements of the system. It does not decrease in an insulated system, according to the second law of thermodynamics (Dodé, 1979, 291).

20. One can obtain chaotic behavior even in apparently noncomplex phenomena that are represented simply by a recursion equation of one variable.

21. The converse is not true, because null correlations do not necessarily imply disorder.

22. This case may be compared to deterministic systems that are sensitive to their initial conditions, whose details on a very small scale are not precisely known. Roulette *simulates* chance by the fact that the number upon which the ball comes to rest is unpredictable even though it was clearly determined by the conditions under which the ball was thrown onto the wheel. A roulette wheel must be constructed with great precision to assure its symmetries.

23. In the case of a continuous distribution, the probability is replaced by a probability density, $f(x_i)$, which when multiplied by the differential of x_i is the probability that x_i falls within the interval $(x_i, x_i + dx_i)$. The ensemble, or set average, is the average in which each discrete value, x_i, is multiplied by its probability, p_i. In the case of a continuous distribution, it is the average at which probability is replaced by $f(x_i)dx_i$.

24. This is true in a probability convergence and even in a nearly sure convergence (Kolmogorov, 1950).

25. In the case of turbulence (see chapters 2 and 3), statistical methods are frequently employed, but numerical calculations are sometimes also performed on the basis of deterministic equations.

26. This is not always necessary, because in certain cases the variation of similarity parameters may not have any significant effect on the result in given domains.

27. This is true only of airplanes flying at speeds less than that of sound. At supersonic speed, another similarity parameter plays a role; that parameter is the Mach number, the ratio of the speed of the plane to the speed of sound in the medium through which the plane is moving.

II
Turbulence in Fluid Mechanics

In this chapter, we discuss the behavior of common gaseous and liquid fluids in the context of classic mechanics and within the scale of human observation. We are interested mainly in turbulent flows and, in particular, those turbulence phenomena that may be regarded as fully determined; we specify the criteria for identifying them and analyze the mechanism by which these phenomena develop their own self-regulation, reach an optimum state, on average, and develop uniform statistical properties. (Characteristics we call, in shorthand, *regulation*, *optimization*, and *homogenization.*)

In the next chapter, we consider the fluid mechanics of the atmosphere and hydrosphere because of their importance as the media of life; but in these spheres of fluid motion we also have to consider other processes, such as changes of phase that pertain to water, radiant energy, and certain chemical reactions. The conditions that characterize determinism, as well as the mechanism and average effects of the homogenization, regularization, and optimization of terrestrial climate, are examined in this context. In the conclusion of this chapter, we compare the behavior of this inanimate medium with that of animate organisms and consider their suitability in a teleonomic arrangement that permits adaptation and support of life in the biosphere.

II.1. Fluids and Turbulent Flows

II.1.1. At the human scale of observation, the states of fluids like air and water have the properties of a continuous medium. Actually, although molecules form a discontinuous medium, human senses perceive only the average effects of their motion.[1] Statistical methods permit the definition and measurement of these effects (Brun et al., 1968; Dodé, 1979).

Fluids are defined by their chemical composition, or by the composition of two components in the case of a mixture of two liquids or two gases, and by the following dependent variables or unknowns: mass per unit volume, or density, ρ; pressure, p; internal energy per unit of mass, e, or absolute temperature, ϑ; the three components u_1, u_2, u_3 (or u_i, where $i = 1, 2, 3$), of the velocity vector; and the concentration, c, of a constituent diluted in a binary mixture. Each fluid is characterized by the values of its coefficients of viscosity, μ, thermal conductivity, λ, mass diffusion, D, and specific heat at constant volume, c_v, and pressure, c_p. These coefficients depend only upon temperature, according to equations known from considerations of statistical mechanics and experiments.

The dependent variables are functions of the independent variables, that is, time, t, and the coordinates x_1, x_2, x_3 (or x_k, where $k = 1, 2, 3$) with respect to the axes of reference associated with Earth. Chemical reactions—phase changes from gas to liquid to solid and back, as well as electromagnetic effects—are for the most part not considered in this chapter.

II.1.2. When we observe the flows of fluids, we distinguish turbulent flows and nonturbulent flows, and transition flows between them. These different types of flow may coexist with only certain zones, being the sites of turbulence, such as wakes, jets, boundary layers along side walls, and thermal plumes. These types of flow may also succeed each other intermittently, particularly in a transition zone.

Because of the transparency of air and water, the differences between the types of flow often pass unperceived. However, at a free surface one can see the vortices that characterize turbulence distributed irregularly in time and space. Swirling movement can also be seen in the atmosphere when it contains dust particles. Turbulent motion is also evident in the wake of a ship, in the course of a rapid stream, and in smoke or leaves blowing about. In the laboratory, turbulent flows may also be observed by optical techniques, such as photography, shadowgraphy, or interferometry.

The frontispiece (see I.3) shows an eddying turbulent flow composed of the wakes of two orthogonal cylinders in a uniform current of water. The red and green colors indicate which fluid has passed close to the cylinders. At first, the wakes preserve distinct eddy structures; then they tend toward complete turbulence. Further downstream, the color of the wakes becomes homogeneous because of turbulent diffusion.

The phenomenon of turbulence is so omnipresent in nature that it may be considered to be the rule in fluid flows, with nonturbulence being the exceptional state. The atmosphere, the medium we live in, is agitated by turbulent flows up to a height of about ten kilometers (within the troposphere). The same phenomenon of turbulence is observed as well in the cosmic domain, in the atmosphere about the other planets, in the sun, and in nebulae.

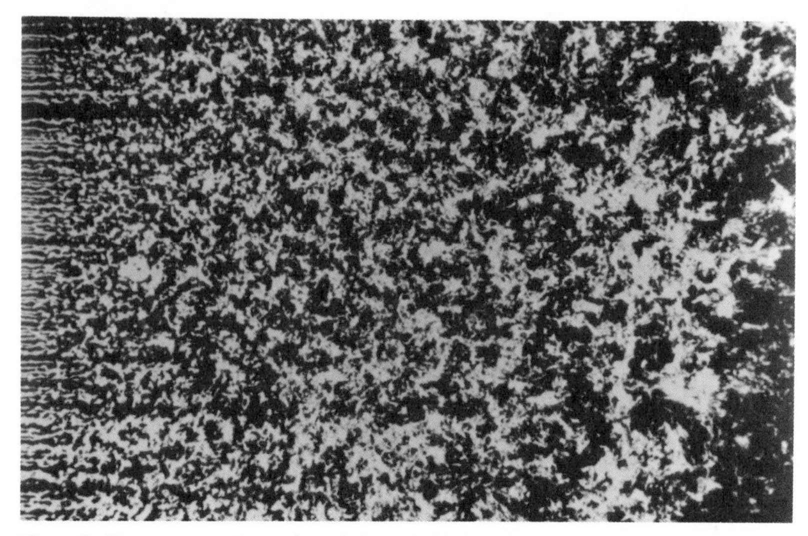

Figure 5. Homogeneous, isotropic turbulence behind a grid. Photograph by Thomas Conke and Hassan Naquib, from Van Dyke, 1982.

In everyday language, a flow is turbulent at a given scale of observation when it is composed of a great many eddies of varied dimensions. In the language of physics, a flow is *turbulent* at a given scale of observation when it is composed of very complex vortical fluctuations having varied time and length scales and, more precisely, when it is composed of *chaotic spatiotemporal fluctuations of vorticity* and of chaotic spatiotemporal fluctuations of the diverse physical properties of the flow, such as velocity, pressure, and temperature (Favre, 1992, 329). This implies that the spatiotemporal correlations between the fluctuations, measured at two points and at two times, tend toward zero when the intervals in space and time between the measurements tend toward infinity. More simply, statistical "memory" has a limited duration. It follows that the spectra in time and space are principally composed of wide bands (see I.4, I.7, and Favre, 1983, 2852–53).

Experiments show (figure 6) that correlations measured in such flows are not null and that statistical memory persists over a long time and at a long distance downstream, meaning that the phenomenon of turbulence is not completely disordered. We shall see, in addition, that it satisfies the criteria for determinism. Thus we can say that turbulence is a *deterministic*, spatiotemporal, vortical chaos.

The chaotic character of turbulent fluctuations excludes periodic and quasi-periodic fluctuations. Their vortically chaotic character also excludes chaotic acoustical fields, because they are not vortical.

By a mixing effect, turbulence in fluid flows, like molecular agitation,

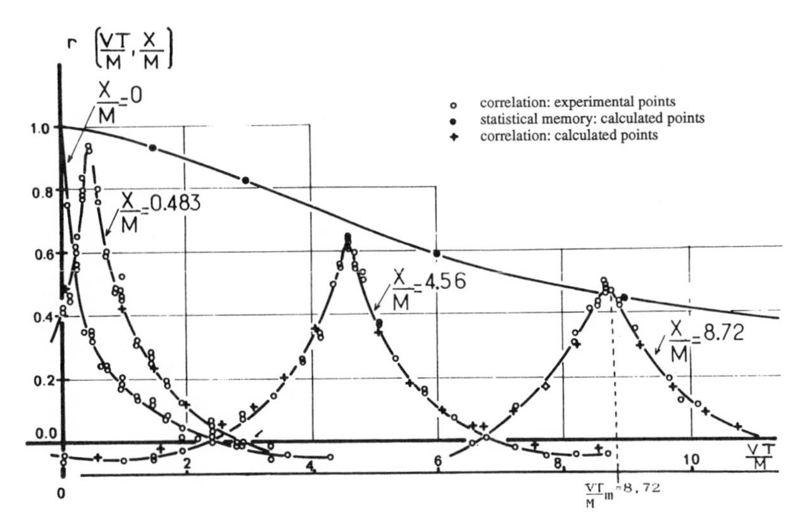

Figure 6. The results of the first measurements and calculations of a spatiotemporal correlation. From Favre, 1965b, 1983; Favre et al, 1976, pt. 3.

diffuses material, heat, momentum, color, and radioactivity—all the transportable properties associated with matter. Because turbulence induces mixing on a much greater scale than molecular diffusion does, the rate of diffusion caused by turbulence generally surpasses by many orders of magnitude the rate of diffusion caused by molecular agitation.

Another effect of turbulence is to increase the velocity of deformation (rate of strain) of a fluid so that turbulent motions render the dissipation of kinetic energy by the action of viscosity more rapid. To take an example, figure 5 uses smoke wires to illustrate the flow behind a grid made of a plate with small holes. This flow is a spatiotemporally determined chaos, rather than a disordered flow. As in figure 6, the correlations are not null. The merging, unstable wakes become turbulent. Downstream, they provide a useful approximation to the idealization of homogeneous and isotropic turbulence from the statistical point of view. We shall see in II.2 that, however complex it may be, the flow is still fully determined.

In figure 6, *r* is the coefficient of spatiotemporal correlation and statistical memory for the turbulent wake of a grid, inserted into an average, straight, and uniform flow: that is, a spatiotemporally deterministic vortical chaos. Spatial separations, *X/M*, are constant; temporal intervals, *VT/M*, are variable. Maximum values for *r*, which are obtained by the optimal temporal interval, VT_m/M, are represented by the curve that envelops the correlations—that is, statistical memory. These results concern a fully established turbulence in the wake of a grid, composed of two orthogonal ranges of rods, placed in a wind tunnel within a flow of air whose average velocity, V, is constant. The edges of the square meshes that compose the grid have a length, M, which serves as the reference for measuring the dimensionless distances, X/M, in the longitudinal dimension of the wind tunnel. The Reynolds number based on that length is 21,500 (Favre, 1965b; 1983).

The coefficient of spatiotemporal correlation, r, is measured between two fluctuations, u_1' and u_1', of the longitudinal component of velocities considered, respectively, at two separated points in space in the direction and the distance, X, and at two moments in time, separated by an interval, T (see I.4). Measurements and calculations are performed for various constant distances, X/M, and with various time intervals, VT/M.

The calculations, which are based upon the spatial correlation and the simple hypothesis that turbulent diffusion is governed by a Gaussian law, agree with experiments up to a first approximation. When distance, X/M, vanishes, we obtain a temporal correlation that decreases rapidly and tends toward zero as the temporal interval, VT/M, increases, and its Fourier transform is a temporal spectrum composed of wide bands of frequencies characteristic of temporal chaos. When VT/M vanishes, one obtains a spatial correlation, which also decreases and tends toward zero as X/M increases, and its Fourier transform is a spatial spectrum composed of a wide band of wave numbers characteristic of spatial chaos.

Let us now consider the general case, where the intervals are both spatial and temporal. For example, for $X/M = 8.72$, r, the coefficient of correlation that nearly vanishes when VT/M is zero, reaches a maximum of 50 percent when the interval reaches an optimal value of VT_m/M. The ratio of distance, X, and optimum time, T_m, defines a statistical celerity, C, for the turbulent field. When the observer is moving with a celerity equal to the celerity of the field, the coefficient of spatiotemporal correlation reaches a maximum, which is the statistical memory, represented by the curve enveloping those maxima whose area defines the duration of statistical memory (Favre, 1983, 2853).

Both spatiotemporal correlations and statistical memory tend consistently toward zero when the intervals of space and time increase. The duration of statistical memory is limited, but it does not vanish. Consequently, this turbulent flow is indeed a spatiotemporal chaos, but it is not disordered chaos (see I.7.4), and as we shall soon see, such fluid flows are completely determined, while in figure 5, the flow appears to be disordered. This flow and those similar to it—that is, those whose walls are geometrically similar and that have the same Reynolds number (see I.9)—are thus spatiotemporally and vortically chaotic but are nevertheless still completely determined.

Experiments performed in boundary layers of fluid flows where the mean velocities have high gradients have yielded qualitatively consistent results (Favre, 1965b, 1983; Favre et al., 1976). However, statistical celerity, C, which is equal to mean velocity, V, when the latter is constant—in figure 6 it is obvious that $X/M = VT_m/M$—may differ from V in the general case, where mean velocity is not uniform. This is explained by the presence of nonlinear terms in the equations for optimum correlations (Favre, 1983, 2854; Favre et al., 1976, 213–15).

Numerous experiments have been performed by other investigators of varied flows (Favre, 1983, 1852): boundary layers, mixing layers, jets, flows in pipes, wakes, and even supersonic flows. These experiments have all yielded results qualitatively consistent with those just described and even quantitatively consistent, by order of magnitude, with regard to statistical memory, when it is considered as a function of the distance, X_1, reduced to the scale of the turbulence, Λ_1, defined as the area under the curve of spatial correlation, and with regard to the intensity of the turbulence,

$$(X_1/\Lambda_1)(\overline{u_1'^2})^{1/2}V^{-1}.$$

Thus the turbulent flows that have been studied and those similar to them are completely determined spatiotemporal vortical chaos.

Indeed, it is reasonable to expect that many natural phenomena, like the turbulence in fluid flows just discussed, fluctuate in space and time and may be considered nondisordered and spatiotemporally chaotic if their spatiotemporal correlations are not null and if either they tend to vanish when the intervals in space and time increase or statistical memories have nonzero but limited durations.

II.2. Turbulent and Nonturbulent Fluid Flows and the Conditions for Determinacy

II.2.1. The behavior of fluids in nonturbulent, transition, and turbulent flows satisfy the conditions of physicomathematical determinism as defined in I.5.2.

II.2.1.1. Let us consider first the general conditions. The physical system is completely defined. It is composed of a fluid defined by its chemical composition or possibly by the chemical composition of two constituents; it is delimited in space by a closed surface. When its local state is observed at the scale of the continuum, it may be represented by the values of a finite number of properties, the unknowns, ρ, p, u_1, u_2, u_3, e or ϑ, and c, which depend upon a finite number of independent variables of space and time, x_1,

x_2, x_3, and t. The physical properties that are particular to each fluid may be expressed by the values of μ, λ, D, c_v, and c_p, which are functions of temperature, ϑ. The complementary, inertial force to which the fluid is subjected, due to the rotation of the frame of reference, is represented by the terms for the Coriolis force, which depend upon the rotation of Earth and the momentum of the fluid.

The system may evolve when it undergoes those transformations that are necessarily governed by the principles and laws of classic mechanics and classic physics and that are expressed in a set of equations, whose number is equal to that of the unknown variables. Thus those transformations are expressed in closed equations. The fundamental principles of physics are (1) conservation of mass, (2) conservation of energy (the first law of thermodynamics), and (3) conservation of momentum (Newton's second law of motion).

In classic mechanics and classic physics, at the scale of the continuum (Howarth, 1956), (1) the principle of conservation of mass is expressed by the equation for density, ρ, and, in the case of a binary mixture, by another equation for the concentration of one of its constituents, c; (2) the principle of conservation of energy may be expressed by the equation for internal energy, e, or for temperature, ϑ; (3) the principle of conservation of momentum is expressed by three equations that correspond to the three components of momentum, ρu_1, ρu_2, and ρu_3. Thermodynamics permits us to describe the state of a gas by an equation relating p, ρ, and e or ϑ for gases, while, for liquids, ρ depends upon ϑ exclusively. Thus we have seven general equations for seven unknowns, ρ, p, u_1, u_2, u_3, e or ϑ, and c. However, these general equations include terms relating to the average effects of molecular diffusion, terms that must be rendered more explicit. These formulas were written according to laws for the behavior of diverse fluids, which reflect the differences among the materials of which they are composed.

A single component of a mixture diffuses within it from a zone of high concentration to a zone of low concentration. Heat is diffused from a zone of higher temperature to a zone of lower temperature. Momentum is diffused from a zone of higher momentum to a zone of lower momentum. These effects of molecular diffusion tend to homogenize the properties related to materials.

The theory that describes the behavior of isotropic fluids with linear viscosity (Howarth, 1956; Brun et al., 1968; Truesdell, 1974; Favre et al., 1976) assumes that both the vectorial mass flux of a component in a mixture and the vectorial heat flux are related, in the former case, to the vectorial gradient of the concentration of that component and, in the latter, to the vectorial gradient of the temperature by ρD, the coefficients of proportionality of the diffusion of the mass, and by λ, the coefficient of thermal conductivity. This theory also assumes that the tensor of molecular stresses—apart from the pressure, which is its isotropic part—is linearly related by the coefficient of viscosity, μ, to the rate-of-strain tensor.[2] The general equations of fluid mechanics thus formulated are called Navier-Stokes equations, which are nonlinear, partial derivative equations.

On the molecular scale, one can also apply the same fundamental principles. When developing molecular dynamics, one takes as a starting point equations for the movement of molecules subject to interactions; then one takes averages to determine the continuum properties of the fluid. But these operations are limited by the capacity of computers, and so one usually uses statistical methods to obtain the averages directly.

The kinetic theory of gases permits, first, a definition of *density* as the mass of molecules in a unit volume and, second, a separation of the average translatory kinetic energy of molecules into two parts, one that represents average motion weighted by mass of molecules and one that represents internal energy. The absolute temperature is proportional to the average kinetic energy of the fluctuating molecular motion. The principle of conservation of momentum applied to fluctuating molecular motion and to collisions that occur among the molecules shows that pressure is equal to two-thirds of the average kinetic energy of the fluctuations (Chapman and Cowling, 1939; Dodé, 1979). The equation for the state of a perfect gas is, as we have already mentioned, a relation among ρ, p, and e or ϑ. This theory also explicitly determines μ, λ, D, c_v, and c_p. The general equations for the continuum state, which relate density, internal energy or temperature, and momentum, are in fact recovered by these statistical methods in formulations identical to those of classic mechanics.

To establish the laws describing diffusion on the basis of molecular motion, Ludwig Boltzmann (Dodé, 1979; Howarth, 1956) supposes that, in a molecular system that has not yet attained its thermodynamic equilibrium, the number of complexions by which that system may attain its equilibrium tends toward a maximum. David Enskog shows that, when Boltzmann's equation is calculated to a first approximation, it describes only pressure, whereas when it is calculated to a second approximation, it also describes mass diffusion, thermal conduction, and stresses of molecular friction (Howarth, 1956), the same laws produced by the methods of classic mechanics. When the calculation is extended to a third approximation, it adds only nearly negligible terms.[3] These laws have been confirmed by experiments up to good approximations. The fundamental principles of classic mechanics and physics have been verified by numerous observations and experiments with ever-increasing precision and have never been called into question.[4]

The number of Navier-Stokes equations is thus equal to the number of unknowns. Thus the general structural conditions for physicomathematical determinism are satisfied in the case of fluids, whether their flows be non-turbulent, turbulent, or in transition between those states. This corresponds to the philosophical concept of necessity, because the behavior of fluids may not violate the principles and laws of mechanics and physics. However, these necessary general conditions are not sufficient for a complete determination of a fluid flow; the particular circumstances of each flow must be taken into account.

In fact, general Navier-Stokes equations apply to all possible flows of common fluids that have the physical properties just described and, thus, to all possible flows of air or water. Now, each flow actually occurs in particular circumstances, which must be precisely described. This is expressed in the mathematical language of physics by the claim that fundamental principles describe the net variations of diverse quantities. These principles are formulated by Navier-Stokes equations, which contain first-order partial derivatives with respect to time and space and second-order derivatives with respect to space. It is well known that the general integrated solution to such equations contains arbitrary constants and arbitrary functions. These constants and functions correspond to contingent factors relative to the particular flow being studied; to express them explicitly, one must take into account the special conditions of realization of that flow.

II.2.1.2. Each fluid flow responds to its own particular circumstances, which are contingent. We have seen that this does not contradict freedom of action, since a person may intervene to modify them (see I.5.2 and I.5.5). These particular circumstances must be susceptible of formulation in a manner compatible with the general Navier-Stokes equation. They include boundary conditions, forces acting upon the system from outside, and the system's initial conditions.

Boundary conditions are always present at the frontiers of the system of any flowing fluid one may be considering, that is, at the closed surface enclosing the flow. These are entirely contingent conditions, since this surface may either be constituted of walls, fixed or movable and impermeable or not to the contained matter, heat, and radiation; or the surface may be constituted of free surfaces that separate two fluids; it may be constituted of virtual surfaces, chosen to complete the closed surface.

By their contact with the fluid, the walls contribute to (or diminish) its velocity and temperature. For instance, the effects of viscous friction appear in the adherence of the fluid to the wall. When a wall is impermeable to the fluid flowing within it, the velocity of the fluid with respect to that wall is null at the contact surface. If, however, the wall is permeable to the fluid then the velocity with respect to the wall will have both tangential components, which are null, and a normal component, which represents the flow across the wall. The temperature of the fluid at the surface of contact is equal to that of the wall because of thermal conduction; and if the wall is impermeable to heat, the derivative of temperature in the direction normal to the wall is null.

At the free surface separating two immobile, immiscible fluids, these conditions require equality of pressure and temperature of the two fluids at the interface. In the typical case of a free surface, such as between water and air, the effect of diffusion of mass (water vapor) also appears as equality between the pressure of saturation of the liquid and the vapor pressure corresponding to concentration at the surface (Brun et al., 1968).

The case of a free surface separating fluids in motion, like the surface of a sea roiled by waves, is much more complicated and thus much more difficult to formulate, but the interactions at the boundary are no less real. To the preceding conditions one must add the effects of inertia — the surface tension of the droplets of liquid suspended in the gas — and of the bubbles of gas within the liquid (Favre and Hasselmann, 1978). When dealing with virtual surfaces, one must choose them so that the boundary conditions are susceptible of a simple formulation.

With regard to forces outside the system acting upon elements of the volume of fluid, we consider here only gravity. Gravity is the result of the attraction of the mass of the fluid by the mass of Earth and of the centrifugal force due to Earth's rotation and is equal to the product of density and the acceleration of gravity, g, which is known. This force is contingent only in special cases, where it may vary or even vanish (weightlessness), as occurs in spacecraft missions and satellites, where gravity is compensated by centrifugal force due to the curve of the trajectory. We might also consider other exterior forces, such as electromagnetic radiation and electromagnetic effects, as contingent conditions.[5]

The state of a system at a given moment, which amounts to initial conditions, is another contingent condition of a particular flow. The memory duration for molecular motion of a fluid with linear viscosity (a Newtonian fluid) is very short with respect to the scale of the continuum, so we describe its state at the initial moment. When the flow is not turbulent, initial conditions are simple and may be easily formulated. When the flow is turbulent from the initial moment, initial conditions are harder to recognize and formulate. One then statistically describes the known properties of the velocity, pressure, density, temperature, and (possibly, in the case of a mixture) concentration of a component.

But for practical reasons — the limits on our ability to measure certain properties and to acquire sufficient data — the description is often cut off at the smaller scales of fluctuation. It follows that precision of prediction decreases with an increase in the period of the prediction from the moment when the initial data were measured (Lorenz, 1969a, 1969b). However, our inability to describe and treat these complex initial conditions in detail does not diminish their objective, physical reality, nor does it affect the behavior of the fluid. This insufficiency of human skill implies nothing with respect to the determinism or indeterminism of the physical phenomenon itself (see I.5 and I.6). It is best to remove this difficulty by choosing the initial moment as the one when the flow is not in transition or a state of turbulence; the initial conditions will then be well known and easily formulated.

II.2.1.3. Once the general, structural conditions are satisfied, in confirmation of the fundamental principles of physics, and the particular circumstances of the case being studied have been formulated, the solutions of the equations are completely determined and, consequently, so are the states of the fluid they represent, for every moment in the future and at every point within the spatiotemporal domain.

Jean Leray (1934) proved the existence of a solution for the Navier-Stokes equation in the sense of a distribution, that is, a generalized function corresponding to mostly arbitrary initial conditions, when a fluid fills a space. This solution exists globally for all times after the initial moment. Under stricter hypotheses about the initial data, this solution has regularity and is unique over a sufficiently short temporal interval. Thus nonturbulent, transitional, and turbulent fluid flows satisfy all the criteria, general and particular, for physicomathematical determinism.

The flows of real fluids are irreversible phenomena because of molecular diffusion, which produces viscous friction, thermal conduction, and mass diffusion. Above all, future states are determined on the basis of the initial state, but for a limited temporal interval.

II.2.2. One may also study the states of fluids by the purely experimental techniques of physical determinism. Such a study is based upon the replicability of the phenomenon (defined in I.5.3) for each experiment performed under the same general structural conditions, that is, it is repeated within the same domain with the same fluid, which of course obeys the same physical laws and is, as well, subject to the same particular boundary conditions, external forces, and initial conditions. If the states and movements of the fluid are in fact replicated for each experiment at every moment subsequent to the initial one and at every point within the domain, the system is considered to fulfill the criteria for experimental determinism up to the limit of the precision of the measurements.

Such verifications of the theory have in fact been performed a great many times for nonturbulent fluid flows. In the case of transition or turbulent flows, experiments are difficult to replicate under identical circumstances, in particular in regard to initial conditions. When possible, it is preferable to choose an initial moment when the flow is not turbulent. It is also difficult to verify in complete detail that transition and turbulent flows are identical. With this objective in mind, delicate measuring methods combined with visualizations and computations of flows still have to be developed.

II.2.3. When we deal with transitional or turbulent flows, we usually treat experimental, physical determinism by statistical methods (see I.5.4). These methods are based on the replicability of the phenomenon up to an average for each series of experiments. The conditions are the same as those of experimental determinism, except that the definitions, the state, the properties, the boundary conditions, the external forces, and the initial conditions of the system are each represented by a statistical average taken for each of a series composed of a great many experiments performed under similar circumstances.

If one recognizes that the average states and average motions of the fluid do in fact replicate for each series composed of a great many experiments, at every moment subsequent to the initial one and at every point within the domain where measurements are taken, the flows, whether nonturbulent, transitional, or turbulent, are considered to fulfill the requirements of experimental, statistical determinism up to the limits of the precision of the measurements. Such statistical verifications have been performed in numerous series of trials, each of which was composed of a great many experiments.

In the case of statistically stationary (and ergodic) flows, that is, those whose statistical properties are independent of time, the ensemble (set) averages and temporal averages are equal. The latter averages are the easiest to take in experiments performed in wind tunnels, hydrodynamic conduits, and even aerohydrodynamic tunnels holding a flow of air above a flow of water, which creates waves (Favre and Hasselman, 1978).

In the domains that have been studied, the results of experiments performed according to these methods have shown that measurements for fluid flows, whether nonturbulent, transitional, or turbulent, replicate themselves up to an average and fulfill the criteria for experimental, statistical determinism, within the limits of precision of measurement.

The replicability of the properties of flows, up to an average, has been verified even for cases where one cannot be sure that the initial conditions have been repeated precisely in all their detail. This is true even for turbulent flows that are very sensitive to initial conditions. In the attraction basin, the phase trajectories that are initially close diverge rapidly, but they subsequently converge toward the attractor. The final, average properties are nearly independent of minor differences in initial conditions (see I.3; Bergé et al., 1984).

In any event, statistical measurements performed in varied types of turbulent flows show that correlations between fluctuations in diverse quantities — like velocity and pressure and possibly temperature, density, and concentration — generally do not vanish. By the method of spatiotemporal correlations, which permits tracking the field of turbulence during its propagation, one even discovers that its statistical memory is quite long (see I.4 and figure 6). Experimental observations also suffice to show that these turbulent flows are not disordered (see I.6), because the diverse properties are not independent. The impression of disorder or chance that we get, subjectively, from the complexity of transition flows and turbulent flows is not objectively justified (see I.6, I.8, and figures 5 and 6).

II.2.4. To conclude, the states and fluid flows in the nonturbulent, transitional, and turbulent cases fulfill the criteria of physicomathematical determinism. In fact, the general laws of mechanics and physics, what we have called the *general conditions*, are necessary and apply to all types of flows, each of which is subject to the special conditions of its particular situation. These special conditions are contingent and even compatible with freedom of human action.

II.3. The Causes of and Criteria for the Existence of Turbulence: Similarity and Transition

When a fluid flow is stable, the perturbations it may acquire are damped, and its state is nonturbulent. When the flow is unstable, the perturbations are amplified, the state is transitory, and it eventually becomes turbulent when it develops very complex fluctuations as a result of nonlinear effects. The diverse types of the flow are predictable on the basis of the specific properties of the fluid and the particular circumstances of each flow — its boundary conditions, its initial conditions, and the exterior forces that act upon it. In principle, these predictions can be made from the solutions of the closed Navier-Stokes equations for fluids (see II.4.1), but they require extremely long calculations.

II.3.1. Considerations of dynamic similarity simplify the problem considerably for each type of flow, that is, for flows that have the same initial conditions, exterior conditions, and boundary conditions — in particular, geometrically similar walls (see I.9; Brun et al., 1968; Monod-Herzen, 1976).[6] These equations may be reduced in nondimensional terms and would then be independent of the units of measurement. Lengths, times, and masses — or even better, lengths, velocities, masses, and temperatures — would then be related to lengths, L, velocities, V, masses, ρ_o, and maximal deviations of temperature, $\Delta\vartheta$, which define the type of flow being considered and individualize the case being studied.

The equations may then be written in the same terms but with dimensionless coefficients — the parameters of similarity — which may assume different numerical values. When these values are equal, the equations and their solutions are identical. This condition is sufficient (but not always necessary) for dynamic similarity, because the ratios of homologous terms are identical. Knowing all about one flow permits knowing about many different flows of the same type. It is sufficient to multiply lengths by L, velocities by V, masses by ρ_o, and temperatures by $\Delta\vartheta$. The trajectories will be geometrically similar, which will obviate turbulent flow being similar to nonturbulent flow. Experimental and theoretical studies may then be restricted to a single flow of a given type, and varying the parameters of similarity will yield all the desired information regarding all similar flows of that type, which is what permits testing upon models.

In the category of flows of a liquid or a gas at a constant temperature, constant density, constant viscosity, and without a free surface (considered together, because their equations of state vanish), a single parameter of similarity intervenes in the dimensionless equations, the Reynolds number, R_e, which is equal to $\rho VL/\mu$. This number represents the ratio of the forces of inertia to the force of viscosity. It is thus sufficient to study a single flow in any such fluid as a function of this parameter. The production of turbulence then occurs by a kinetic process, and its damping is due to viscous friction.

Osborne Reynolds was the first to utilize this parameter, in 1883, to study experimentally flows in cylindrical pipes. In such cases, L represents the diameter of the pipe, V the average velocity, ρ the density of the fluid, and μ its viscosity. He then rendered the flow visible by colored filaments. In the nonturbulent case, the filaments remain straight and separated, in the transition case they twist, and in the turbulent case they assume very complex fluctuating forms and diffuse very quickly, coloring the entire mass of the flow homogeneously (see frontispiece). Reynolds also observed that the loss of pressure along the length of the pipe increases suddenly with the appearance of turbulence. Finally, he discovered that the passage from one state of the flow to another depends only upon the Reynolds number for the given initial state. The flow is nonturbulent if the Reynolds number is less than a critical stability value of about 2,000. When the Reynolds number increases, the flow passes through a phase of instability; the flow becomes turbulent when the Reynolds number reaches a transition point. Turbulence then continues, whatever values R_e may assume that are greater than the critical value.

The result has a general meaning because, for all types of flows that may be studied, the kinetic turbulent state occurs once the Reynolds number exceeds the critical values of first instability and then transition. These values are somewhat smaller for jets and wakes than for flows in pipes; they are intermediate for flows along a wall, but they remain within the same order of magnitude.[7]

Observations of the transition of flows from a nonturbulent to a kinetic, turbulent state detect the diverse processes of instability in which fluctuations develop very rapidly under the nonlinear effects of inertial forces. On the contrary, however, when the production of turbulence ceases, turbulence decays very gradually through the dissipation of kinetic energy by viscous friction.

In the most general case, the density is not constant. The flow may vary and may even become turbulent, either because of variations in temperature or in the mixture's concentration of gases or liquids with different densities or because of the turbulence in high-speed flows when the Mach number exceeds 0.5. In such cases, turbulence is still produced by instability, but it develops according to two processes. The first process is the kinetic process just described, which occurs whether the density is constant or variable. This is the kinetic production

$$-\overline{\rho u_i' u_k'}\, \delta \tilde{u}_i / \delta x_k,$$

which is proportional to the gradient of the mass-weighted average velocity, $\tilde{u}_i$ (shear flow) multiplied by the correlation between the fluctuation of momentum and that of a velocity component.[8]

The second process, which appears in the statistical equations (Favre et al., 1976, pt. 2; Favre, 1983, 2858), is due to the enthalpic production,

$$-\overline{u_i' \frac{\delta p}{\delta x_i}},$$

whose major part is proportional to the mean pressure gradient multiplied by the correlation between density and velocity in the same direction. This occurs when the density is fluctuating and when its correlation with velocity is significant, producing a mean pressure gradient. The pressure gradient

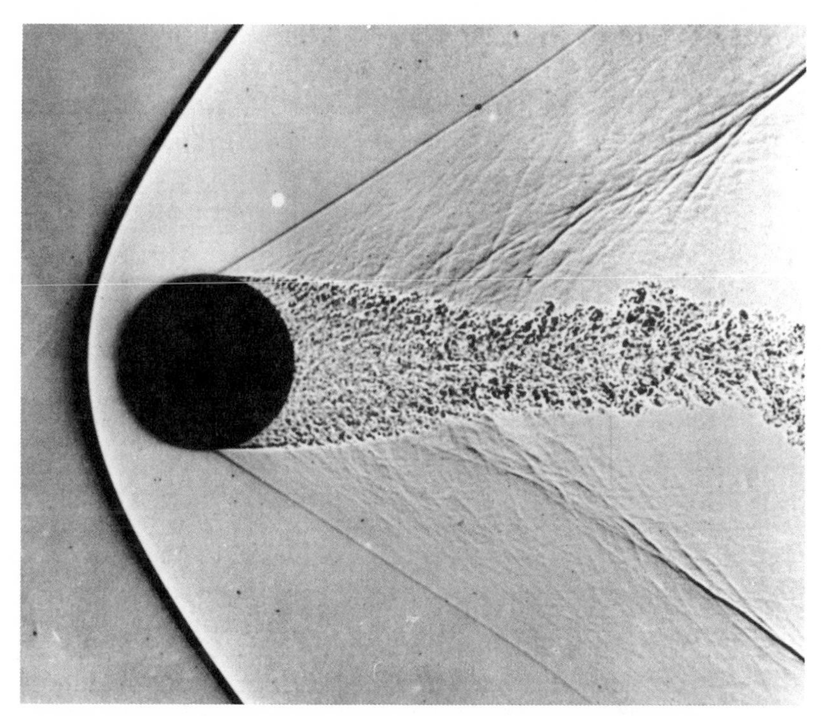

Figure 7. A sphere flying at supersonic speed (Mach 1.53), producing shock waves and a turbulent wake. Photograph by A. C. Charters, from Van Dyke, 1982.

may be due to Archimedes' force in the gravitational field, in which case the enthalpic production is expressed by $-g\,\overline{\rho' u_1'}$, which indicates buoyancy by thermal-turbulent (or mass-turbulent) convection (u_1' is the vertical component of velocity fluctuation). An analogous turbulence may also occur in the pressure gradient due to the inertial forces in high-speed flows (supersonic flows or high centrifugation) by enthalpic production.

Kinetic turbulence develops according to the above indications when the Reynolds number exceeds the critical values in the zones with high gradients of average velocity, such as pipes, canals, wakes, jets, boundary layers in the neighborhood of walls, free surfaces, and interfaces between fluids moving at different speeds (see the frontispiece and figures 4 and 5). Figure 7 is a shadowgraph of a half-inch sphere caught in free flight through air at Mach 1.53. The flow is supersonic ahead of, and subsonic behind, the part of the bow shock wave that is ahead of the sphere and wrapped around its surface to 45°. At about 90°, the nonturbulent boundary layer at the sphere's surface separates through an oblique shock wave and quickly generates a turbulent wake. This wake produces weak disturbances, which merge into the second shock wave.

Turbulent convection develops in zones with small mean velocity and unstable stratification, that is, a rising particle has a lower density than the ambient fluid, and its ascending movement is maintained by Archimedes' force. On the other hand, if the stratification of the fluid is stable, vertical currents are dampened. The stratification of density may be induced by the vertical gradient of potential temperature or by the concentration of a component of the mixture, such as humidity in the air or salinity in the sea.

The parameter of similarity that describes thermal convection above a horizontal surface is the Rayleigh number, R_a, discovered in 1916, which equals $g\beta\Delta\vartheta L^3\rho_o/\mu\alpha$. In this expression, g is the gravitational acceleration, β is the coefficient of the thermal expansion of the fluid, α its thermal diffusion, $\lambda/\rho c_p$, while $\Delta\vartheta$ represents characteristic temperature deviation. When the average velocity is low or vanishes, transition is accomplished by successive bifurcations, where motions appear to be organized in the well-known Rayleigh-Bénard cells or rolls (Bénard, 1901; see figure 8), after which these motions become chaotic and turbulent. They occur as soon as the Rayleigh number exceeds the critical values, the least of which is 700.

In the more general case, where both kinetic turbulence and convective turbulence may coexist, it is more convenient to use the Reynolds number and the dimensionless Richardson number. The Richardson number for the flux, R_f, defines a stratification of the density that may be stable or unstable and that is equal to the quotient of convective (enthalpic) production of turbulence by kinetic production of turbulence, taking the opposite sign by convention. When convective production of turbulence is null, R_f is null, and the only pertinent criterion is the Reynolds number. When both types of turbulence exist, since the kinetic contribution is often positive, the Richardson number is negative. It is positive if this kinetic turbulence is, on the contrary, absorbed into the potential energy of the gravitational field. For large negative values of R_f, convective turbulence is preponderant.

In the atmosphere, convective turbulence may be produced by rising plumes, which can reach an altitude of about ten kilometers (within the troposphere).[9] Stratification is stable at higher altitudes (within the stratosphere). Kinetic and convective turbulence often coexist in the atmosphere. Kinetic turbulence develops in the ocean within narrow layers about the interface with the atmosphere and within the major ocean currents. Convective turbulence produced by thermal effects and by differences of salinity can sometimes penetrate to great depths. Thermal transport and the transport of mass are not exclusively due to turbulence; there is also a mechanism for their transfer by the mean motion of a fluid.

Thus the causes of turbulence in fluid flows are the instabilities within which fluctuations develop. They are related to boundary conditions, to initial conditions, and to exterior forces, principally gravity. Turbulence occurs when the critical values of the Reynolds, Rayleigh, or Richardson parameters are exceeded. The Reynolds parameter defines kinetic turbulence within zones where average velocities have highly differing values, such as boundary layers at walls or free surfaces, wakes, and jets. The other two parameters define convective turbulence within a gravitational field in zones of unstable density stratification. This unstable stratification may be due to sharp differences in temperature, which give rise to thermal convective turbulence, or it may be due to the differences of concentration in a mixture of fluids whose densities differ, giving rise to mass convective turbulence with a diffusion into the mixture. These processes may coexist.

In geophysical fluid flows and in phenomena created by human intervention, the critical values of the parameters of similarity are often exceeded, which is why turbulence is the rule and nonturbulence the exception.

II.3.2. The various types of flows and the various types of transition from the nonturbulent state to the turbulent state are so numerous and so diverse that their classification is not yet complete. Although all these phenomena are completely determined — they obey the appropriate equations — the theories that scientists have succeeded in elaborating concern only certain types of transition.

Theories of hydrodynamic stability have been developed for nearly parallel flows when the perturbations to which they are subjected are sufficiently small to permit the omission of the nonlinear terms of the equations, at least initially (Schlichting, 1955; Betchov and Criminale, 1967). One calculates the critical Reynolds number below which the flow would be stable. Above that number, one determines the neutral curves that constitute the limits of the frequencies of the unstable fluctuations. When the latter occur, nonlinear interaction effects appear and may lead to a turbulent state.

In the case of nearly parallel flows in pipes or in the boundary layers along the walls of a conduit, as well as in the case of flows between concentric cylinders when the exterior cylinder is rotated, the first appearance of turbulence occurs as Emmons spots (Emmons, 1951; Hinze, 1975, 605), whose successive bifurcations cannot be detected. These spots are irregularly distributed in space and time, producing intermittent zones of nonturbulence and turbulence. The zones of turbulence develop and merge when the Reynolds number for transition is achieved, and then the state becomes entirely turbulent. For other types of flows, such as Couette's, between two cylinders, when the interior cylinder is set into rotation, as well as in numerous cases of turbulence due to convection, bifurcations occur in the course of transition.

Figure 8. Buoyancy-driven convection rolls. Photograph by Oertel and Kirchartz, from Van Dyke, 1982.

During the last several years, the qualitative study of the types of transition has been greatly furthered by the theory of dynamic systems, due to the work of Ruelle (1978) and Ruelle and Takens (1971) on strange attractors (see also Swinney and Gollub, 1981). These writers suggest that the chaotic state may be reached after a small number of bifurcations.

The fundamental example is that of Rayleigh-Bénard thermoconvective instabilities. A fluid is placed between two horizontal surfaces that are good conductors of heat. The temperature of the lower surface exceeds that of the superior surface by $\Delta\vartheta$. The Rayleigh parameter of similarity, which is proportional to $\Delta\vartheta$, also plays the role of a bifurcation parameter. When the values of this parameter are low, the velocity of the flow is null, and the transfer of heat is accomplished exclusively by conduction. When the first critical value is reached, there occurs a bifurcation that corresponds to the appearance of stationary vortices, or rolls, on parallel horizontal axes. They revolve about each other clockwise and counterclockwise, alternately, with a minimum of friction between adjacent rolls. This produces, alternately, a rising flow, which is warmed by the lower surface, and a descending flow, which is cooled by the upper surface. The flux of heat by convection is added to the flux of heat by conduction, producing a stronger heat flux by a transfer mechanism, which seems to be optimal. For increasing values of the parameter, other instabilities appear, and the process leads to a state of thermoconvective turbulence. In figure 8, differential interferograms show side views of the convective instability of silicone oil, in a rectangular box of the relative dimensions 10 : 4 : 1, heated from below. At the top is the classical Rayleigh-Bénard (Bénard, 1901) situation: uniform heating produces rolls parallel to the shorter side. In the middle photograph, the temperature difference — and hence amplitude of motion — increases from right to left. At the bottom, the box is rotating about a vertical axis.

These transition mechanisms may be interpreted according to the "theory of dynamical, dissipative systems" (Swinney and Gollub, 1981). The Navier-Stokes equations for fluids (see II.4.1) applied to the particular circumstances of each type of flow are simplified by rather strong hypotheses, which reduce them and transform them into three ordinary Lorenz differential equations (Lorenz, 1963, 135; Bergé et al., 1984, 131), whose solutions may be calculated by computer.[10] The variables X, Y, and Z are exclusively functions of time; X is proportional to the intensity of the convective motion, Y is proportional to the difference of temperature between the rising and descending currents, and Z is proportional to the difference in the vertical temperature profile with respect to the linear profile. The Prandtl number, P_r, is the ratio of the molecular diffusion of the momentum to heat and is particular to the given fluid; r is a control parameter; b is positive and figures in the term $-(P_r + b + 1)$. This term is equal to the speed of contraction (divergence) of the volume in the phase-space, which thus tends toward zero by *dissipative* effects. Thus the trajectories that were initially included within a given volume ultimately *converge toward the attractor*, which has a null volume.

According to Edward Lorenz, his equations can yield realistic results when the Rayleigh number is slightly supercritical but not in the case of strong convection, because of the simplifying hypotheses that were adopted. Figure 9 shows the attractor he obtains when he chooses the values $P_r = 10$, $b = 8/3$, and $r = 28$ (Lorenz, 1963, 137). Thus r is slightly higher than the critical number, 24.74, that corresponds to the second bifurcation — in fact, it is an inverse — beyond which a chaotic state appears, with a "strange attractor."

This figure shows projections, first on the XY plane and then on the YZ plane, of the phase-space for the parts of several trajectories that correspond in calculation to iterations 1,400–1,900 (written 14–19). The states of stable convection are shown by points C and C'. It is evident that X and Y are often the same signs (strong correlation), in which case they correspond to a warm fluid rising or a cold fluid descending, both of which carry heat by convection. Starting from diverse initial conditions, the trajectories form spirals about point C, then about point C', then again in the neighborhood of C; they are finally confined by the attractor. This strange attractor was the first discovered; contraction is rapid and the system is highly dissipative. This strange attractor has a *fractal* dimension of 2.06.[11]

When one continues the calculations for values of the parameter of bifurcation that lie between 30 and 214, one obtains a very complex diagram

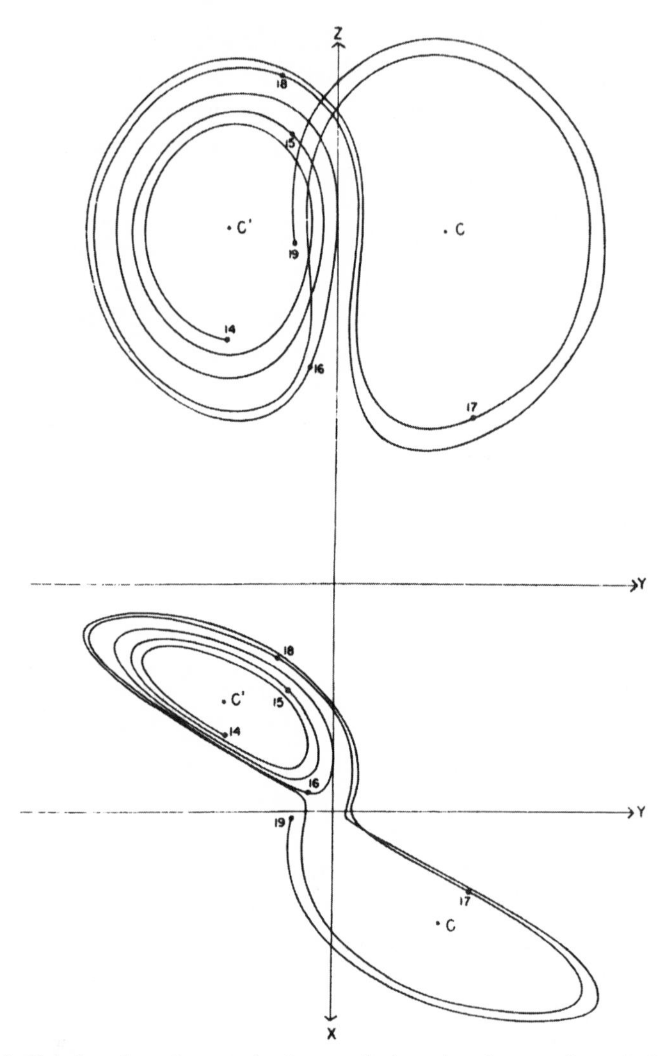

Figure 9. Turbulent thermal convection between horizontal surfaces, or Lorenz's "strange attractor." From Lorenz, 1963.

of solutions for the equations, with an alternation between chaotic states and periodic states, inverse cascades, subharmonics, and intermittent zones (Bergé et al., 1984).

II.3.3. With regard to flows within which turbulence is thoroughly established, qualitative and even quantitative studies are most often performed by statistical methods (Hinze, 1975; Favre et al., 1976, pts. 2 and 3). As we shall shortly see (in II.4.2), the statistical equations are not closed because of the nonlinearity of the equations for fluid mechanics, which produces new unknown terms in the average operation. It is thus necessary to establish closure models, and there are theories of turbulence adapted to the various types of flow (Hinze, 1975) and based on hypotheses for which we give several classic examples (see II.5). Currently, attempts are being made to associate the theory of dynamic systems with these hypotheses by retaining the large-scale structures and making models for separated smaller eddies (Aubry et al., 1988).

A study of turbulence touches upon several disciplines: fluid mechanics, thermodynamics, astrophysics, chemistry, and biophysics. It has applications to such varied fields as aeronautics, astronautics, hydraulics, thermal and nuclear generation of energy, and manufacturing techniques in chemistry, agriculture, and medicine. Its results are pertinent to meteorological-oceanographic predictions and pollution prevention.

II.4. The Mechanisms and Regulating Mean Effects of Turbulent Flows

Fluid flows carry with them all their transportable properties, such as matter, sensible heat, latent heat (internal energy), and momentum. When the flow is turbulent, it is customary when using statistical techniques to separate the transfer of these properties into those effectuated by the average motions of the fluid and those due to turbulent fluctuations. The latter induce a mixture whose average effect is turbulent *diffusion.*

Just as molecular agitation produces an average diffusion effect, the stage or stages of turbulent fluctuations also produce average diffusion effects. And just as molecular diffusion produces a homogeneity of fluid properties, turbulent diffusion produces average *homogenization* effects with regard to the concentration of the constituents of the mixture, internal energy, and momentum.

The effects of turbulent diffusion are often of a higher order of magnitude than the effects of molecular diffusion. In addition, by increasing the speed of deformations in the movement of a fluid, turbulence produces an average effect that strongly increases the viscous dissipation of kinetic energy in the form of heat, which in turn dampens the turbulence. Reynolds's experiments show that the turbulent diffusion of mass is much stronger than the molecular diffusion of mass. The colored filaments in a nonturbulent fluid flow diffuse only very slowly when they are exclusively effected by molecular agitation, while turbulence diffuses them very rapidly, as can be seen as their color spreads out homogeneously in the fluid. Numerous experiments have confirmed the general nature of this property (see the frontispiece).

Turbulent flows of fluids in the atmosphere and hydrosphere (see III.2) tend to homogenize the properties of those media by the transfer effects of both average motion and turbulent diffusion. It is these mechanisms that maintain the constant composition of dry air, even though it is, in fact, a mixture of gases of different densities. These mechanisms diffuse humidity in the air and salinity in the oceans, they transfer great quantities of sensible heat within the air and the oceans, and they transfer the latent heat of water vapor in the air and salt in the oceans, thus acting to limit differences in temperatures and concentrations within those media. Thus turbulent and molecular diffusion constitutes a regulation mechanism for a fluid medium. In such cases, the kinetic energy of turbulent fluctuations and its dissipation are weak in comparison with the thermal, and sometimes chemical, energy that turbulence diffuses, which corresponds to an optimization of its average effects.

These phenomena may be explained quite precisely, because they can be given mathematical expression and also may be studied experimentally. We shall now try to do this, using a minimum of formulas and referring the reader to the literature for complementary information.

II.4.1. The general equations for fluids express the fundamental principles of mechanics and physics in the language of mathematics: (1) conservation of mass, (2) conservation of total energy, and (3) the rate of variation in momentum being equal to the forces applied to the fluid. The equations for the state and behavior of a fluid (see II.2.1.1), which represent the average effects of molecular agitation, depend upon the same principles, expressed in statistical terms, at the scale of the continuum.

To derive the general equations, one defines a control volume, Ω, within the domain occupied by the fluid and limited by a closed and fixed surface, S. One then expresses the balance of variations for each of the transferable properties of the fluid that is found at every instant within that volume and flowing across that surface. The fundamental properties whose variations are governed by the above three principles, considered by unit volume, are mass and internal plus kinetic energy, $\rho e + \frac{1}{2}\rho u_i u_i$, the three components of momentum, ρu_i $(i = 1, 2, 3)$.[12]

To simplify, and to illustrate the analogy among the various transfer mechanisms, we may group the properties (as well as ρc, which is the mass per unit volume of a constituent diluted in a binary mixture) and denote them by ρw, which represents a scalar quantity or a vector component (Favre et al., 1976, pt. 2).

The balance of variation for the transferable quantities, ρw, is thus determined within volume Ω by taking the rate of variation with respect to time to be a consequence of the following efficient causes: the flux of transfers across surface S due to the motion of the fluid, the flux, H_k, across S due to molecular diffusion, and F, the rate of the creation or destruction of ρw per unit of time and volume. In habitual practice, the surface integral of the two fluxes is transformed into a volume integral of the divergences of these fluxes. The equation is integrodifferential within volume Ω. Since the balance of variation is valid in an arbitrary control volume, it is also valid in the smallest volume defined by the scale of observation at the continuous-medium scale and may be written in terms of a local equation in partial derivatives.

In the general equation, the local balance of variation of a transferable quantity, ρw, is thus obtained (Hinze, 1975, 15), with the first term representing the rate of local variation within the time of ρw and the second term including terms describing the transfer by the motion of the fluid and the transfer by molecular diffusion.[13] These transfers are cumulative and depend upon mechanisms of the same nature, the divergence of their fluxes. The third term, F, represents the local rate of creation or destruction of ρw. The general equations for fluids are obtained by expressing ρw, H_k, and F in a manner consistent with fundamental principles.

The equation for the conservation of mass is obtained by replacing ρw by ρ. If the fluid is homogeneous, the flux, H_k, of the mass by molecular diffusion is zero, and F is also zero, because there can be no creation or destruction of mass. In the case of a binary, inhomogeneous mixture, the equation for c, the concentration of one of the constituents, is obtained by replacing ρw by ρc, the mass of that diluted constituent. The molecular diffusion, H_k, is determined by the law for its behavior, and F is zero, because the mass of each constituent is conserved.

The three equations for the momentum are derived by replacing ρw in each of them by ρu_i and H_k by the average molecular stresses, that is, by pressure and viscous friction, f_{ik} (see II.2.1.1). F must be equal to the components of the force of gravity, ρg_i, and to the components of the Coriolis force. The transfers of a momentum by the motion of the fluid itself and by the stresses of molecular friction, $-f_{ik}$, are cumulative and depend upon the same mechanism, the divergence of the fluxes,

$$\frac{\partial}{\partial x_k}(\rho u_i u_k - f_{ik}).$$

The equation for internal and kinetic energy without a change in state is derived by replacing ρw with $\rho e + ½\rho u_i u_i$ and replacing H_k with the flux of the internal energy, the sensible and latent heat due to molecular diffusion (Favre et al., 1976, pt. 2). F, the rate of creation or destruction of energy, would be null because energy is conserved if it is total energy. But it is necessary to take into account the potential energy due to gravitational attraction, $\rho u_i g$.

As for the Coriolis force, it does not figure in total energy variations because it is orthogonal to the velocity and thus makes no contribution. By subtracting from this equation the equation for the kinetic energy obtained by a combination of the momentum equations, one obtains the equation for internal energy in which appear the dissipation of kinetic energy as heat and the power that corresponds to the work of the pressure as the volume changes. Within the equation for internal energy, the transfers of energy by the movement of the fluid and by molecular diffusion (flux E_k) are cumulative and depend upon the same mechanism, the divergence of the fluxes,

$$\frac{\partial}{\partial x_k}(\rho e u_k + E_k).$$

When the terms that represent molecular diffusion (whose laws are described in II.2.1.1, where mass diffusion is denoted by D, viscosity by μ, and heat conductivity by λ) are introduced, the equations obtained are the Navier-Stokes equations for fluid flows. Thermodynamics permits a determination of the state equation for perfect gases (an equation applicable to air, even humid air, with a modified constant, R) and the state equation for liquids, which shows that ρ depends upon temperature only insofar as expansion is a function of temperature.[14]

II.4.2. The flows within which turbulence is well established are so complex that one cannot learn all their details and chooses to limit oneself to determining, by statistical methods, their average properties, the only properties of any practical use. On the contrary, the direct method consists of solving in detail the closed Navier-Stokes equations for the particular circumstances of the flows and then calculating the statistical averages of the solutions to determine the average quantities and the average effects of the turbulence, which can then be applied to the particular cases being studied. But it is only in very simple cases that direct solutions for turbulent flows may be obtained, with all the details, by mathematical analysis, and only in very limited cases may then be obtained by numerical calculation.

The statistical method commonly used consists of taking, a priori, statistical averages for the terms of the Navier-Stokes equations. The statistical equations thus obtained relate to the averages of the properties being studied. But the operations for obtaining the averages introduce new unknown terms, with correlations between fluctuations (see I.4) when they are applied to nonlinear terms. These equations are not closed. One can write equations for these unknowns, but they would, in turn, introduce other unknowns, and the process would diverge. The object of the theory of turbulence is therefore to propose as many additional equations as there are new unknowns in order to close the equations. To describe nature, linear equations are the exception and nonlinear equations the rule. Turbulence in a fluid flow, which is essentially a nonlinear phenomenon, is a good example.

To explain the average effects of turbulence qualitatively, it is not necessary to resolve the equations. One may consider the physical significance of the terms of the statistical equations, especially of those that express correlations between fluctuations, that is, the mean effects of turbulence. As for the quantitative approximations of these average effects, they may be obtained from the theoretical solutions and from the statistical measurements that may be performed during the course of the experiments.

II.4.2.1. The study of the turbulent flows of fluids was begun toward the end of the nineteenth century by Joseph Boussinesq and Osborne Reynolds, who limited their work to cases where density, temperature, and viscosity are constant and the fluid is homogeneous. The only turbulence that develops in such a case is kinetic turbulence, since turbulent convection may not occur, as the fluctuation of ρ is zero. The cases that fall into this category are in fact numerous, including hydraulics and aerodynamics when the speed does not exceed half the speed of sound, that is, a Mach number less than 0.5.

The unknowns are reduced to four, p, u_1, u_2, and u_3. The same is true for the Navier-Stokes equation. The equations for internal energy and for the state of the fluid vanish. There is, therefore, a complete analogy between liquids and gases and also a complete similitude when the particular circumstances are similar and the Reynolds parameters are equal, so experiments may be performed with either gases or liquids.

Reynolds discovered the statistical equations that bear his name for fluids with constant density (Hinze, 1975). For such fluids, he separated the velocity components, u_k, into average velocities, $\overline{u}_k$, and fluctuations, u_k'', whose averages thus vanish; he did the same for pressures. Then he took the averages of the equations, themselves. Statistical equations have the same form as general equations for the linear terms, and it suffices to replace pressure, velocity, and viscous friction by their averages. But for terms representing transfer of momentum, which are not linear, the operation of taking averages introduces new terms for the flux, $\rho\overline{u_i''u_k''}$, with correlations between the fluctuations of velocities. These terms are the components of the Reynolds tensor of turbulence stresses. The effect of turbulent diffusion is added to the effect of molecular diffusion and to the average motion transfer.

The mechanisms of turbulent and molecular diffusion are similar, and their divergences are additive. They tend to homogenize momentum. Theory holds and experiments confirm that, in most cases, total stresses, which are composed of turbulence stresses and the average viscous stresses, are much higher than the exclusively viscous stresses that develop in nonturbulent flows; thus the fluxes of the momentum are greatly increased. Viscous friction acts to dampen the turbulent fluctuations of the velocity, u_k'', of the fluid. At a coarser scale than that of turbulence, in most cases (except, for instance, in a secondary pipe flow), total average stresses act to dampen fluctuations in the average velocity of fluids, $\overline{u_k}$.

This agrees with the fact that reiterated experiments under the same average conditions converge toward the same average values according to the methods of statistical determinism (see II.2.3).

The fact that total stresses are greatly increased by the average effect of turbulence was already clear in Reynolds's experiments on flows in pipes. For a given difference in pressure between inlet and outlet, the average velocity of the delivery of the fluid is considerably decreased when the flow is turbulent. We shall see that turbulence has a similar effect in streams, where it decreases the average velocity of the water flow; and thus turbulence constitutes another regulating mechanism of the mean flow at the higher level of observation.

If one considers a body immersed in a fluid in motion, and if one calls the neighborhood of that body the boundary layer, the velocity of the fluid decreases to the vanishing point at the surface of contact wih that body due to the adherence of the fluid to that surface because of its viscosity. The total stress of viscosity and turbulence exerts a force upon the surface, a force that begins to act at the leading edge of that object—and that is expressed by the dimensionless coefficient of total average friction. This coefficient is much higher in turbulent flows than in nonturbulent flows, in a ratio that increases with the Reynolds number (Rebuffet, 1968). If this number is defined as a function of L, the distance from the leading edge, the coefficient of friction is increased by turbulence in the proportions of 1.8, 7, or 36, depending upon whether the Reynolds number is 10^5, 10^7, or 10^9, respectively.

In flows without constraining surfaces—like jets, wakes, and mixing layers moving with different velocities—turbulence stresses are much stronger than viscous friction. According to the statistical equations for kinetic energy obtained by combining the equation of motion, the kinetic production of turbulence takes the form described in II.3. By increasing the rate of strain, turbulence also has the effect of intensifying the dissipation of energy, which dampens the fluctuations of velocities.

II.4.2.2. The study of the more general case of turbulent flows with a variable and even fluctuating density has been undertaken only recently. Variations in density may be caused by variations in temperature, variations in the concentration of the mixture of constituents with different masses, and variations in pressure at high speeds (higher than Mach 0.5). Variations in temperature may also lead to variations in viscosity, in thermal conductivity, in the diffusivity of mass, and to a lesser degree, in the two specific heats.

Kinetic turbulence and enthalpic turbulence may in fact develop either simultaneously or separately. This situation is characteristic of such significant domains in fluid mechanics treated by thermodynamics as the atmosphere, the oceans, and supersonic flows. One deals not only with the transfer of the heat, mass, and momentum of a fluid but also with the transfer of the mass of a constituent diluted in a binary mixture and with the transfer of internal energy, represented by sensible heat and latent heat, to which one may add chemical energy (discussed in chapter III). One recovers the seven unknowns and the seven general equations of Navier-Stokes. But since the state equations are different, the analogy and similitude between gases and liquids are no longer complete. Experiments must be performed in wind tunnels for gases, in hydrodynamic conduits for liquids, or in aerohydrodynamic tunnels for gases and liquids separated by a free surface.

To establish the statistical equations for turbulent flows of fluids in this case, some writers have retained Reynolds's method of decomposition of variables for fluids of constant density. They separate all the variables, ρ, p, u_1, u_2, u_3, e or ϑ, and c, into their averages and their fluctuations with zero averages. But the decomposition of transportable quantities, ρu_k, ρe, and ρc, per unit volume, which is thus required, is such that their fluctuations do not have a vanishing average. The operation of taking an average of the general equation for the balance of a transportable quantity introduces four new terms with double or triple correlation.[15] The formulation of statistical equations is exact, but they are not simple; and the physical meaning of their terms is not clear because they include many new terms with correlations.

Then there arises a question of which physical quantities are suitable to decompose into their averages and fluctuations. It would seem logical to retain the fundamental physical properties whose variations per unit volume are determined by the three principles of mechanics and physics. These transferable quantities are density, ρ, momentum components, ρu_k, internal energy, ρe, and mass per unit volume of the constituent diluted in the mixture, ρc.

We shall see in IV.1 and IV.2 that the invariance of a physical theory under a transformation group determines which are the suitable physical variables. Momentum corresponds to the translation group in space and energy corresponds to the translation group in time. It is not velocity that is significant but its product with mass. These variables are used in classic statistical mechanics and in relativistic mechanics.

The method we adopted to develop a formulation of the statistical equations for fluids with variable densities (Favre, 1958, 1965a, 1983, 1992; Favre et al., 1976, pt. 2) is the decomposition of fundamental properties, that is, transferable quantities, ρw, into averages, $\overline{\rho w}$, and fluctuations, $(\rho w)'$, whose averages vanish. For density, ρ, we recover the decomposition into $\overline{\rho}$ and ρ'. Quantities, w, are decomposed into a nonfluctuating part, $\tilde{w}$, which must be invariant in the average, and a fluctuation, w'. It follows that $\overline{\rho w} = \overline{\rho}\tilde{w}$ and that $\overline{\rho w'} = 0$. The variables, $\tilde{w}$, are mass-weighted averages for velocities, $\tilde{u}_k$, internal energy, $\tilde{e}$, temperature, $\tilde{\vartheta}$, concentration, $\tilde{c}$, and entropy, $\tilde{s}$.[16] The statistical equation for the balance of a transferable quantity, ρw, may then be written in a more concise form (compared to the equation written by Reynolds's method in note 15), and these terms have the simplest physical significance, which permits a clear interpretation of their mean effects because it includes only one term with correlation, the turbulent diffusion term.[17]

If one compares the equation in note 17 with the general equation for the balance of variations of a transferable quantity in note 13, one sees that the terms have the same form if one replaces ρ by its average, $\overline{\rho}$, replaces the variables w and u_k by their mass-weighted averages, $\tilde{w}$ and $\tilde{u}_k$, and adds the divergence of the mean flux transferred by the turbulent fluctuations, $\overline{\rho w' u'_k}$ (the turbulent diffusion), to the divergence of the mean fluxes of the ρw's transferred by average motion, $\overline{\rho}\tilde{w}\tilde{u}_k$, and molecular agitation, $\overline{H}_k$.

The first term in this statistical equation for the average balance of ρw represents rate of local variation with respect to time for $\overline{\rho w}$. The second term expresses the average effects of the transfer of ρw by the divergence of the fluxes that is due both to the average motion weighted by mass and to turbulent and molecular diffusion. The third term is the local mean rate for the creation or destruction of ρw per unit of volume and time, $\overline{F}$. The transfers accomplished by the average motion weighted by the mass of the fluid, by molecular diffusion, and by turbulent diffusion are cumulative. Their mechanisms are of the same type: the divergence of their fluxes.

The statistical equations for fluids with variable density may be obtained by developing w, H_k, and F as one does for the general equation. The average transfer effects for turbulent fluctuations are represented by terms for diffusion: Reynolds's stresses, $\overline{\rho u'_i u'_k}$, for the momentum components; $\overline{\rho e' u'_k}$ for internal energy; and $\overline{\rho c' u'_k}$ for concentration. The state equation for perfect gases assumes the simple form, $\overline{p} = R\overline{\rho}\tilde{\vartheta}$. In any event, the combination of equations for momentum leads to the statistical equation for the kinetic energy of turbulence, in which the kinetic and enthalpic, or convective, productions[18] (see II.3) figure, as well as the dissipation of energy (Rayleigh's function; see Favre et al., 1976, pt. 2).

Theory and experiments show that, in most cases, the fluxes that correspond to turbulent and molecular diffusion are much more significant than the exclusively molecular diffusion that occurs in a nonturbulent flow. Turbulence intensifies the tendency toward the homogenization of transferable properties—concentrations, momentum, sensible heat, latent heat—all of which tend to limit the deviations of the average properties of the flow. The effects of turbulence also increase the dissipation of kinetic energy into heat produced by means of viscous friction. This has a damping effect upon turbulence.

The kinetic energy of turbulent fluctuations and its dissipation are often weak in comparison with the internal energy of sensible and latent heat and with the chemical energy they diffuse. In general, turbulent flows have homogenizing, regulating, and optimizing mean effects, which can be observed in the atmosphere and the oceans (see chapter III) at higher levels of observation, as well as in physical theories (chapter IV) in a teleonomic arrangement.

II.5. Some Classic Models for Turbulence

As we have seen, the statistical equations for turbulent flows are not closed. The averaging of nonlinear terms produces new, unknown terms, divergences of $\overline{\rho w' u_k'}$, such as the Reynolds stresses, $\overline{\rho u_i' u_k'}$. Writing the equations for these terms introduces correlations of higher order, like $\overline{\rho w' u_k' u_j'}$, which are more numerous than the new equations, and so on.

There is no general statistical theory of turbulence, but some theories attempt to solve the closure problem by modeling the unknown terms for each type of flow by means of hypotheses inspired by experimental results. These theories are falsifiable, since their predictions may be compared with experimental results.

Certain classic hypotheses have led to the main statistical theories for fluids with constant density (Hinze, 1975; Favre et al., 1976, pt. 3). Boussinesq assumed in 1877 that, like viscous stresses, turbulent stresses are directly proportional to the mean velocity gradient, and he introduced the idea of eddy viscosity, a scalar value (Schlichting, 1955). Objections to this hypothesis are that, as a general rule, a constant value may not be expected and that eddy viscosity should be a tensor (Hinze, 1975).

Despite these objections, Boussinesq's assumption does lead to significant results for many shear flows. Under this assumption, eddy viscosity is added to molecular viscosity, and since eddy viscosity is generally much greater than molecular viscosity, the resulting efficient viscosity is markedly increased by turbulence. If we consider a Reynolds number defined in terms of efficient viscosity, it would be much smaller than the Reynolds number defined in terms of molecular viscosity alone. This provides a heuristic explanation for the damping of the fluctuations of mean velocity.

Sir Geoffrey Taylor in 1915 and, independently, Ludwig Prandtl in 1925 assumed that eddy viscosity is equal to the product of mixing length and suitable velocity, and they modeled it by means of considerations of dimensional analysis (Schlichting, 1955; Hinze, 1975, chap. 5). Subsequently, their theories were extended to the turbulent transfer of heat and mass by means of ideas regarding the eddy conductivity of heat and the eddy diffusivity of mass. Like efficient eddy viscosity, efficient eddy conductivity and diffusivity are strongly increased by turbulence. Other models have been proposed that depend upon statistical equations for the transport of the kinetic energy of turbulence, for its dissipation, and for the Reynolds stresses. There have been improvements and extensions in the applications of these models. Despite objections to the underlying hypotheses, it appears possible to obtain from these phenomenological theories a satisfactory description of the mean properties of such turbulent flows as boundary layers, wakes, and jets. Many further applications to engineering have been discovered in recent decades.

Taylor (1935) and Theodor von Karman and Th. Howarth (1938; see Hinze, 1975; Favre et al., 1976, pt. 3) propose statistical theories of isotropic and homgeneous turbulence in constant density fluids. These theories apply to the simplest type of turbulent flow, since no preference for any specific direction is assumed. A minimum number of quantities and relations is necessary to describe the flow's structure and behavior, so it has been studied most thoroughly, both theoretically and experimentally. But the assumed homogeneity excludes a gradient of mean velocity and the production of turbulence; thus the type of turbulence with which they deal decays because of the dissipation of energy, and it cannot be stationary.

Andrei Kolmogorov (1941) introduces a statistical theory of local isotropy. It was known that the kinetic energy of turbulence may be described by a three-dimensional energy spectrum function, $E(k,t)$, of wave number, k, and time, t, and that, in case of high Reynolds numbers, the energy of large eddies is transferred through nonlinear interactions to smaller eddies. Finally, the smallest eddies lose their energy by viscous dissipation into heat according to an energy cascade.

This inspired Kolmogorov's first hypothesis, which holds that, at sufficiently high Reynolds numbers (for instance, in the atmosphere), there is a range of high wave numbers where the turbulence is statistically in equilibrium and is uniquely determined by dissipation, ϵ, and viscosity, υ (Hinze, 1975; Favre et al., 1976, pt. 3). This state of equilibrium is universal, because such turbulence is independent of particular conditions; and any change in the length scale, η, and the velocity scale, v, can be attributed only to the effects of ϵ and υ.

By means of dimensional analysis, Kolmogorov defined those two scales such that the Reynolds number with respect to this length, η, and this velocity, v, is equal to one.[19] The wave number, k_d, above which the viscous effects become very strong, will be of the same order as $1/\eta$ and will be defined by it. On the other side of the equilibrium range is the wave number, k_e, that marks the range of the eddies in which most of the energy is held. The lower wave numbers correspond to the largest eddies.

For the equilibrium range, $k_e \ll k \ll k_d$, Kolmogorov proposed his second hypothesis. If the Reynolds number is infinitely large, the energy spectrum within the equilibrium range is independent of viscosity, υ, and is in fact solely determined by a single parameter, ϵ. Since transfer of energy is the dominant factor within this subrange, it is called the *inertial subrange.* The form of the spectrum in this subrange is

$$E(k,t) = A\epsilon^{2/3}k^{-5/3};$$

it is called the *Kolmogorov spectrum law* and is well confirmed experimentally.

In 1962, Kolmogorov (1962a, 1962b) reformulated his theory to take into account the intermittency observed in the dissipation of turbulence (see Hinze, 1975; Favre et al., 1976, pt. 3). According to his third hypothesis, the logarithm of the fluctuation of dissipation follows a normal Gaussian law. This theory is consistent with experimental measurements, up to a first approximation, for many flows at very high Reynolds numbers—notably in geophysics, in atmosphere-ocean boundary layers, and in astrophysics observations.

These theories and their corresponding observations show that among the notable mean effects of turbulence developing at a given scale are the damping of fluctuations of mean quantities for shear flows and, for flows with large Reynolds numbers, local statistical equilibrium in the inertial range, both of which exemplify a tendency toward a stable final state.

Just as molecular agitation is a regulating mechanism for the properties of the medium at the continuum scale, turbulent fluctuations are a regulating mechanism for the mean properties at the upper scale of observation. In the following chapter, we consider an important example: the interaction of the atmosphere with the hydrosphere over Earth's surface.

Notes

1. We should bear in mind that our senses cannot distinguish details smaller than 0.01 mm. and 0.001 second, and the techniques developed in fluid mechanics yield thresholds of observation on the order of 10^{-3} mm. and 10^{-6} second. But for air, at normal temperature and pressure, there are 27 million molecules in a cube 10^{-3} mm. on a side, and the average interval between their collisions is 10^{-10} second.

2. In the case where the density is variable, there exists a second coefficient of viscosity. One usually expresses it as a function of the first coefficient using the Stokes formula.

3. An exception is the case of flow-through supersonic shock waves.

4. When speeds approach the speed of light, these laws are generalized by relativistic mechanics, which we discuss in chapter IV.

5. The equations are then completed by Maxwell's equations for electric and magnetic fields and by Ohm's law. Together with the boundary and initial conditions, they lead to a well-posed problem (Moreau, 1990) that satisfies the conditions for determinism.

6. Gases and liquids must be considered separately, because the equations for their states have different forms.

7. In the Reynolds experiment, when all the units of measurement are in the CGS (centimeters, grams, seconds) system, the ratio, ρ/μ, is close to 100 for water, and the critical value of instability is 100 VL for pipes 1 cm. in diameter; the critical velocity is therefore about 20 cm./sec. For air, ρ/μ equals approximately 7, and the critical value for instability is $7VL$. For a pipe 10 cm. in diameter, the critical velocity is 28 cm./sec. In the atmosphere, if we take L to be the thickness of the boundary layer above ground level, which is around 1,000 m., the value of this parameter is $7 \times 10^5 V$. For the case of a wind of 3.6 km./hr., the Reynolds number reaches 7×10^7. We see that if the wind is not nearly null, the boundary layer of atmosphere is in a state of kinetic turbulence.

8. The viniculum $(\overline{\ldots})$ is taken to mean the ensemble, or set, average, while the symbol $(\widetilde{\ldots})$ is taken to mean the ensemble, or set, mass-weighted average, for example, $\tilde{w} = \overline{\rho w}/\overline{\rho}$ (see II.4.2). The summation convention with respect to repeated indexes, i_i, k_i, is accepted ($i = 1, 2, 3$; $k = 1, 2, 3$).

9. The rising plumes may be amplified by the release of latent heat from the condensation of water vapor in the clouds (see III.2).

10. The Lorenz differential equations are:

$$dX/dt = P_r(Y - X); \qquad dY/dt = -XZ + rX - Y; \qquad dZ/dt = XY - bZ.$$

11. See Bergé et al., 1984, 155. The Poincaré section of the Lorenz attractor for $Z = r - I$ shows two segments, which could lead us to say that it is a surface

(dimension 2); but crossing the different sections, one sees that this attractor is strange; it comprises many layers separated by a vacuum; it is neither a surface nor a volume; its dimension is fractal: $D = 2{,}06$. We can consider a set of points and recover this set by (hyper) cubes of size E. The minimum number of cubes for that purpose being $N(\mathrm{E})$, the dimension is the limit when E tends to zero, if:

$$D = \lim_{\mathrm{E} \to 0} \frac{ln\, N(\mathrm{E})}{ln\, (1/\mathrm{E})}.$$

12. In fact, total energy also includes potential energy, $\rho u_i g$, which corresponds to the work of gravity, which figures in the net variation.

13. The Hinze equation may be written,

$$\frac{\delta}{\delta t}(\rho w) + \frac{\delta}{\delta x_k}(\rho w u_k + H_k) + F = 0.$$

14. The state equation for perfect gases is $p = R\rho\vartheta$.

15. The statistical equation for the balance of a transferable quantity ρw by the Reynolds average method is (Favre, 1969; 1992; Favre et al., 1976)

$$\frac{\delta}{\delta t}(\bar{\rho}\,\bar{w} + \overline{\rho'' w''}) + \frac{\delta}{\delta x_k}(\bar{\rho}\,\bar{w}\,\bar{u}_k + \bar{\rho}\,\overline{w'' u_k''} + \overline{\rho'' u_k''}\,\bar{w} + \overline{\rho'' w''}\,\bar{u}_k$$
$$+ \overline{\rho'' w'' u_k''} + \bar{H}_k) + \bar{F} = 0.$$

16. We define the velocity with components, $u_k = \tilde{u}_k + u_k'$, where $\tilde{u}_k$ is the mass-weighted average velocity.

17. By the mass-weighted average method, the statistical equation for the balance of a transferable quantity may be written (Favre, 1992, eq. 7):

$$\frac{\delta}{\delta t}(\bar{\rho}\tilde{w}) + \frac{\delta}{\delta x_k}(\bar{\rho}\tilde{w}\tilde{u}_k + \overline{\rho w' u_k'} + \bar{H}_k) + \bar{F} = 0.$$

18. The mean value of the enthalpic production of turbulence reads:

$$\overline{-u_i'\,(\delta p/\delta x_i)} = -(\delta/\delta x_i)(\overline{u_i' p'}) + \overline{p'\,(\delta u_i'/\delta x_i)} + (\overline{\rho' u_i'}/\bar{\rho})(\delta\bar{p}/\delta x_i)$$

(Favre, 1983, eq. 10.) The first term on the right-hand side is usually analyzed with the diffusion terms. The second term on the right-hand side is usually small for flows at moderate Mach numbers. The last term on this side may be important when density fluctuations are correlated with the velocity and when the mean pressure gradient is strong. When the pressure gradient is due to gravity, it produces *buoyancy.* Correspondingly, when it is due to inertia—for instance, in supersonic flows—it produces analogous effects.

19. The Kolmogorov scales are

$$\eta = \left(\frac{\upsilon^3}{\epsilon}\right)^{1/4}; v = (\upsilon\epsilon)^{1/4}; \text{ with } v\eta/\upsilon = 1, \text{ and } k_d = 1/\eta.$$

Kinematic viscosity is $\upsilon = \mu/\rho$.

III

The Atmosphere and the Hydrosphere: The Media of Life

We first offer a brief description of the properties of the hydrosphere and atmosphere (Bureau des Longitudes, 1984; Chabreuil and Chabreuil, 1979), the media within which terrestrial life developed. Some explanations of the mechanisms that govern the behavior of the atmospheric-hydrospheric system are then presented. We employ the ideas presented in the preceding chapters regarding turbulent fluid flows, with additional material concerning the phase changes of water from vapor to liquid to solid, transfers of radiant energy, and certain chemical reactions. The behavior of this system is considered from the point of view of its determinacy and the mean regulatory and optimizing mean effects of the terrestrial climate.

III.1. The Properties of the Hydrosphere and the Atmosphere

III.1.1. Our planet rotates on its axis in an ellipsoid. Its equatorial radius is 6,378 km., and its polar radius is 6,357 km. Its mass is about 5.974×10^{24} kg., and its period of rotation of twenty-four hours fluctuates by several milliseconds. Its satellite, the moon, whose mass is eighty-one times less than that of Earth, orbits it in a little more than twenty-seven days at a mean distance of 340,000 km. The center of gravity of the Earth-moon system describes an elliptical orbit about the sun at a distance that varies between 147×10^6 km. in January and 152×10^6 km. in July. Earth's equatorial plane is inclined by 23°27′ with respect to its orbit about the sun.

Proceeding from the center of Earth, one distinguishes internal and external cores, a mantle, a lithosphere that lends the body rigidity and is composed of plates in very slow movement (several centimeters a year) over Earth's crust, the hydrosphere, and the atmosphere and its environment.

According to the *Encyclopédie scientifique de l'univers* (Bureau des Longitudes, 1984),

> the Earth has the unique characteristic among all the planets of the solar system of having, near its surface, light, air, and water. The energy of solar radiation, which is principally collected by the continental and marine terrestrial surfaces, is transferred and exchanged into thermal or mechanical forms, particularly across the marine surfaces; it is then transported, sometimes for great distances, by winds and ocean currents. These exchanges, transfers, and transportations *determine* the range of physical conditions within the fluid environment of the Earth, [its] atmosphere, and oceans and *adjust* it within the narrow range within which *life is possible.* The means of these conditions constitute the atmospheric or oceanic *climate.* The energy received from the sun is eventually dissipated into space. (p. 145)

III.1.2. The hydrosphere consists of water in the oceans, seas, lakes, and rivers: it is principally liquid, which is nearly incompressible, but is also ice, in the form of continental and marine glaciers, and snow. It is a thin layer in comparison with the terrestrial radius and, because of gravity, lies upon the continental surfaces and the marine depths. Leaving aside the closed seas, the world's ocean occupies 71 percent of the surface of the planet and has an average depth of 3.8 km. Its volume is about $1{,}370 \times 10^6$ km.^{3}, or 97 percent of the water known to be contained within the hydrosphere. Its thermal capacity is about $1{,}370 \times 10^{18}$ kcal.K^{-1} ($5{,}7 \times 10^{21}$ J.K^{-1}).

Atmospheric pressure is about 1.013 mbar, or 1,013 hPa, at sea level. Within a volume of water, the pressure increases by 1 kg./cm.^{2} for each 10 m. of depth. The average salinity of marine waters is 3.472 percent of the weight of water, and the average temperature is 3.52°C. Water vapor in the atmosphere derives from evaporation, 85 percent from marine surfaces and 15 percent from continental surfaces. The surface temperature of the oceans varies daily by 0.2°C and annually in the intermediate latitudes by 4°C, at the equator by 1°C, and at the higher latitudes by 9–10°C. In the Atlantic Ocean, surface temperatures vary between 27°C at the equator and 0°C at the highest latitudes, while salinity varies between 3 and 3.7 percent for the same latitudes. The warming of the seas, which is largely caused by the absorption of sunlight (which extends to a depth of about 50 m.), varies daily and annually. Most of this variation occurs within the upper layer of the ocean down to the thermocline, where the vertical temperature gradient is maximal.

A second permanent thermocline does not vary with the season. Its depth, between 100 m. and 200 m. below the surface, increases with latitude. Ocean currents, which carry heat from lower latitudes to higher ones, are caused by the friction of wind upon the surface of the water and by the difference between the heat received from solar radiation at the equator and the lesser amount absorbed at higher latitudes. As a result of the winter cooling of the surface by the action of cold continental winds, the surface water sinks and then causes deeper currents along the meridians. The speed of the vertical descent of the water may then reach 10 cm./sec. down to depths of several thousand meters. These turbulent currents, both at the surface of the oceans and within their depths, tend to homogenize temperatures at a planetary scale, to reduce maximum deviations of temperature between lower and higher latitudes, and to serve as climatic regulators.

Turbulence is also a regulatory mechanism for flows of water within the continents (see II.4.2). Continental and marine glaciers constitute a reservoir of sweet water, which also plays a hydrothermically regulating role.

III.1.3. A layer of air, the atmosphere, envelops the surface of the continents and the oceans, whose total mass is about 5×10^{18} kg. and whose thermal capacity is about 1.2×10^{18} kcal.K.$^{-1}$. It exerts a pressure upon continental and water surfaces at sea level as a result of its weight. Since pressure is zero at the top of the atmosphere and since it is compressible, density increases with pressure; it decreases with altitude at a rate that also depends upon its temperature. While 99 percent of the mass of the atmosphere lies between sea level and an altitude of 30 km., 99.9 percent is between sea level and 60 km., and only the last 0.1 percent is above 60 km.

It is customary to distinguish several parts of the atmosphere. The troposphere is the most important part of the atmosphere because it contains 80 percent of its mass, most of its energy, its cloud cover, and its precipitation. Often unstable and generally turbulent, the troposphere is in contact with the continental surfaces and the oceans, with which it exchanges mass and energy. The bottom of the troposphere is the surface of the continents and the ocean; its upper boundary is the tropopause, where the temperature is at its lowest. The average altitude of the tropopause varies from about 7 km. at the poles, where the temperature is −45°C, to 12 km. at the intermediate latitudes, where the temperature is about −55°C, and to 17 km. at the equator, where the temperature is the lowest, −85°C.

The lower stratosphere is above the tropopause, up to a height of 30 km., within which the vertical temperature gradient either nearly vanishes or is positive. This part of the stratosphere is generally stable. At higher altitudes, toward 50 km., there is a "warm" zone of 0°C. Toward 100 km., there is another cold zone, of −80°C. The average free motion of molecules approaches 1 cm.

Beyond the stratosphere is the ionosphere and then the magnetosphere, a transition zone toward the interplanetary void. The Van Allen radiation belts are situated at an altitude of 5,000 km. We restrict ourselves here to descriptions of the troposphere and lower stratosphere, up to an altitude of 30 km., that is, the meteorological atmosphere.

By volume, the composition of the atmosphere is 78 percent nitrogen and 21 percent oxygen, or 99 percent atmospheric gases. These proportions remain constant because of ecological equilibrium and because of molecular and turbulent diffusion, which tends to homogenize this mixture even though the densities of the constituents are different. There are also rare gases like argon, neon, helium, krypton, xenon, and hydrogen, whose proportions are of the order of 10^{-3}. Furthermore, atmospheric air contains other constituents in varying proportions: water vapor (0–7 percent), carbon dioxide (0.01–0.1 percent), sulfur dioxide (0–0.001 percent), and ozone (0–0.00001 percent).

Were the air immobile, the heaviest constituent, carbon dioxide, would form a layer inimical to life several meters thick above the surface of Earth, but turbulent winds diffuse it throughout a substantial altitude. Ozone (triatomic oxygen) exists in an apparently negligible quantity but still plays a very important role. In fact, it is distributed throughout the atmosphere below 60 km. from the surface, with a maximal concentration at a height of around 30 km. It totally absorbs ultraviolet solar radiation, that is, radiation with wave lengths between 0.24 and 0.31 μm—which, were it abundant, would destroy life. We must also mention the danger from certain human activities that decompose the ozone; nitrogen-rich fertilizers release nitrous oxide into the atmosphere, and refrigerants and the inert propellant gases of aerosol sprays are diffused to heights reaching the stratosphere. The absence of the protective layer of ozone may have such biological effects as the inhibition of photosynthesis and the production or erythema and cancers (Bureau des Longitudes, 1984). Variations in carbon dioxide concentration are induced by natural combustion, photosynthesis at the surface of the planet, and the absorption and emission of that gas by the oceans. However, the various kinds of combustion due to human activity, to brush and forest fires, and to the combustion of fossil fuels are capable of increasing its concentration significantly.

As for the variable concentration of water in the air, it is due to evaporation, condensation, and precipitation. The saturation pressure of water in the air, h, depends upon temperature and varies considerably, from 0.2 mbar at -40°C to 12 mbar at $+40$°C. The mass of water, by comparison with that of air, is, according to the state equation, $5h/8p$. Thus when the temperature drops by 10°C, it can contain only half as much water vapor, which leads to a condensation of 3.75 gr. of water. Condensation releases a significant amount of heat because of the very high value of the latent heat in the vaporization of water, that is, 585 cal./gr., which is among the highest for any of the common fluids. The evaporation and condensation of water are of great importance for transfers of energy from the oceans and continents to the atmosphere and, therefore, have a regulating effect upon the climate.

The atmosphere also contains clouds, another result of the condensation and freezing of water vapor. Clouds are limited almost exclusively to the troposphere. The fact that the pressure of water vapor reaches a saturation value, *h*, is not sufficient to induce condensation; particles of condensation are also necessary. The pressure that effectively leads to condensation increases with the curvature of the surface of the liquid phase, in a proportion that may reach 300 percent when the radius of that curvature is 10^{-6} mm., thus inducing significant delays in the phase change. However, that pressure is reduced by 20 percent by the salinity of the particles of saturated condensation. These particles are crystals of marine salts, which come from the evaporation of marine sprays, mineral or vegetal particles, or dust diffused by turbulence up to the tropopause. This explains why condensation is abundant within the troposphere and nearly negligible in the stratosphere and why fogs are common near oceans, especially near the surface of the water. Cloudy air behaves dynamically like dry air because the proportion of the mass of water vapor to that of the air is very small, on the order of one-thousandth.

Only a very small part of the energy of the ocean and atmosphere comes from the lithosphere, the solid crust; for the most part, it comes from solar radiation, which induces horizontal winds and turbulent convective currents. These motions transport very large quantities of sensible and latent heat both vertically and horizontally. Within the atmosphere, the vertical transport of thermal energy operates mainly through turbulent and molecular diffusion produced by winds and through turbulent convection, which may occur up to the height of the tropopause. Horizontal transport of energy is produced by great, turbulent atmospheric currents and meteorological perturbations, which occur at a scale of several thousand kilometers. This process guarantees a meridional turbulent transport of thermal energy quite as significant as the oceanic transport of thermal energy; it tends to even out the extremes of temperatures on the planetary scale and constitutes a climatic regularizing mechanism.

It is evident that the energy of solar radiation is the motor that induces thermal and mechanical exchanges in the hydrosphere and the atmosphere. The flux, which arrives perpendicularly to a surface outside the atmosphere, averages 1.36 kw./m.^{2}. The power received by Earth is thus the product of this flux by its perpendicular section, or 17×10^{13} kw. It is maximal at the equator and nearly null at the poles, but because of the inclination of the equator to the ecliptic, it varies seasonally at every point. Earth's rotation also produces a diurnal variation at every point on its surface. This flux also varies annually as a function of the inverse of the square of the distance from the sun, or around 7 percent, to the profit of the northern hemisphere, where this phenomenon somewhat attenuates the effect of the seasons.

Incident solar radiation is constituted of short electromagnetic waves, within a range of 0.2 to 10 μm, principally in the form of light. Part of that radiation is absorbed by atmosphere and clouds; another part is absorbed by Earth's surface; still another part is reflected back upon the clouds, the ground, the oceans, and the glaciers; and the rest is radiated back into space by Earth. This energy is thus transformed: a higher proportion of wave lengths are long (between 5 and 25 μm), and entropy is increased, as required by the second law of thermodynamics.

Atmospheric gases absorb very little of the solar radiation within the range of visible wave lengths. "This remarkable transparence of the window coincides with the wave lengths at which most of the solar energy is emitted" (Bureau des Longitudes, 1984, 148). On the contrary, infrared radiation is absorbed mainly by water vapor and (to a lesser extent) by carbon dioxide and clouds, thus producing the greenhouse effect in the troposphere. This is yet another mechanism for the regulation of temperature, subject to the condition that human interference has not overly increased carbon dioxide concentration in the atmosphere.

The thermal contrast between oceans and continents has repercussions in the atmosphere. The penetration of the energy of solar radiation into land masses is limited by the nature of their surfaces and the vegetation that covers them. Penetration averages 1 m. for diurnal effects and 15 m. for annual effects. At a scale of 12,000 to 500,000 years, the effects of ice ages can be detected to a depth of about a thousand m. In contrast, heat coming from Earth does not raise temperatures by more than 0.01°–0.02°, except in volcanic regions. Average monthly air temperatures near the surface of the oceans are around 30°C in lower latitudes and 0°C in higher latitudes, but on the continents temperatures vary with the season, from 35°C in the lower latitudes to −60°C in the higher latitudes. By comparison, the moon's temperature varies between 100°C and −250°C, Mars's between 27°C and −130°C, Mercury's between 400°C and −200°C, and Venus's between 470°C and 446°C. On Earth, in the intermediate latitudes, solar heat varies less than it does at the higher latitudes, and the dominant winds come from the west. These two phenomena temper Earth's climate, principally at the western continental coasts. The temperatures on Earth favor life.

III.1.4. The biosphere is situated principally in the interface between the atmosphere and the continents and oceans; in the oceans, it is limited mainly to the zone where light penetrates, or the *euphotic layer*, which has a maximum depth of 100 m. The euphotic layer is warm in all seasons at lower latitudes and in the summer at higher latitudes. The chemical composition of the ocean is practically constant. "The fact that very few living marine organisms would be capable of living in seawater that is twice as salty supports the concept of an average ocean that has been in dynamic equilibrium for a very long time" (ibid., 275).

The principal constituents of the oceans — sodium, chlorine, magnesium, and potassium, all of which play important roles in the biological cycle — are present in nearly constant quantities. As for oxygen, abundant in the atmosphere, it dissolves in water — and its quantity is restored by, particularly, the effect of waves lifted by winds (whose turbulent breaking traps air bubbles) and by biochemical processes.

The depth of the biosphere extends only several meters into the Earth. It extends several tens of meters into the atmosphere for vegetation and several hundred meters for animal life. It contains a very large number of organisms, whose mass is, however, small in comparison to that of the atmosphere and hydrosphere. The quantities of chemical and radiant energy absorbed and, after certain transformations, emitted by the biosphere are small in comparison to the amount of energy introduced into the atmosphere and hydrosphere. But the vegetal covering of the continents plays a role in the thermal exchanges between ground and atmosphere. Biological cycles are in interaction with their medium. Regarding the physicochemical characteristics essential to living organisms and their milieu, Lawrence Henderson (1921) remarks:

> Water and carbonic acid seem to be the *primordial constituent* substances of the medium. Their importance is due to the fact that their properties, like those of the elements that constitute them, hydrogen, carbon, and oxygen, and like those of the compounds that can be formed out of them by successions of chemical transformations, constitute *maxima*. For example, the heat of vaporization, the heat of formation, the solvent capacity, surface tension, etc. . . . for water. The barrier between air and water, the stabilization of neutrality for carbon dioxide, the number, variety, and complexity of the chemical compounds of the three elements. . . . These *maxima* and even more, their coordination, lead necessarily to a large number of *optima* in the natural conditions of life, such as the stability of the temperature of Earth, the intensity of the circulation of all the elements, the penetration of water and carbon dioxide into the ground, the stability of the reaction of the natural waters and organisms, the intensity and variety of the metabolic processes, etc. . . . This is a convergence, the cooperation of a great number of diverse properties that lends them their full value. A noticeable change

> in any one of these properties would lead to an elimination of the possibility of any organic evolution, however minor, such as that recognized by biologists. . . .
>
> The coincidences that one observes in the properties of these three elements, and that are absolutely required in order for there to be any real evolution, like the coincidences that Laplace discovered in the solar system, are much too numerous to be results of chance. The properties in question appear as though they were not themselves affected by the evolutionary processes. Consequently, I was led to the conclusion that the properties of the elements demonstrate a teleological *arrangement.* (1–3)[1]

We would say now, *teleonomical arrangement.*

III.1.5. The overall energy on Earth is in equilibrium: Earth emits as much radiant energy toward space as it receives from the sun. The complex phenomena that intervene in the course of these exchanges have an ultimate regulatory and optimal effect, the maintenance of a constant terrestrial climate at the scale of decades and millenia. Since the last ice age, about 10,000 years ago, variation in average global temperature has been around 1°C.

This regulation in energy has also been found on nearby celestial bodies. But Earth is remarkable for its hydrosphere and atmosphere, in which fluid motion ensures both vertical and horizontal turbulent transport of mass and energy. We are thus led to a similar conclusion about this aspect of Earth that Henderson proposed for physicochemical processes. The behavior of the inanimate media, the atmosphere and hydrosphere, depends upon regulatory mechanisms that happen to depend largely on the same conditions as those necessary for life on Earth. Indeed, even when natural catastrophes destroy life locally and during limited periods, life on Earth is not completely destroyed. It is rather a question of the global optimization of the possible states that are mutually compatible, without which life could not perpetuate itself, develop, and evolve. This is what lends this behavior its teleonomic character.

Below, we explain some of the mechanisms that govern this behavior, drawing upon our remarks in II.2, the physical theories discussed in chapter IV, and the biological theory discussed in chapter V.

III.2. The Mechanisms of the Atmospheric-Hydrospheric System: Determinism, Regulation, and Optimization

III.2.1. The extent to which the fluid atmospheric-hydrospheric system may be considered to be deterministic (see the definitions in II.2) is now considered, subject to a few simplifications.

III.2.1.1. The general conditions for physicomathematical determinism which apply to the atmospheric-hydrospheric system must be specified for a domain limited by Earth's crust and by a surface, adopted by convention, situated 30 km. above sea level. This system therefore includes the troposphere and the lower stratosphere — the meteorological atmosphere as well as the oceans and seas that cover the 70 percent of land under water. It contains two layers that are quite thin with respect to Earth's radius and that are separated by the surface of the oceans, across which exchanges of energy and matter occur. The waters of the oceans are not uniformly salty. Within the lower stratosphere, air contains water vapor but not condensation, while in the troposphere, air contains water vapor and significant quantities of condensation in the form of clouds and mists composed of drops of liquid, ice crystals, and sometimes precipitation. Since the total mass of water contained in the air is on the order of a few parts per thousand, the physical properties of air are not significantly affected when no phase change for water occurs. However, when such phase changes as condensation, evaporation, freezing, and melting occur, they release or absorb significant quantities of sensible heat by the transformations of their latent heat.

The choice of reference axes related to the rotation of Earth has the effect of adding to the force of inertia the complementary Coriolis force, which is, together with gravity, one of the dominant forces in the atmosphere and hydrosphere. At the human scale of observation, this medium appears to be continuous. Observations are also made at the meteorological-oceanological scales throughout the planetary domain.

The physical properties of the atmosphere and the ocean, in the zones within which no phase change occurs, are completely represented by seven unknowns: $\rho, p, u_1, u_2, u_3, e$ or ϑ, and c. These properties depend upon space variables, x_1, x_2, and x_3, and upon time, t. Concentration, c, is that of water vapor in the air or salinity in water. The specific properties (of both air and water) of viscosity, μ, heat conductivity, λ, and mass diffusivity, D, are given as functions of temperature, while specific heats, c_v and c_p, and the latent heat of water vaporization are constant. A determination of internal energy, or enthalpy, entails temperature, θ, because, in the absence of phase change, one can separate the equations for internal energy into an equation for sensible heat and an equation for latent heat, since the latent heat in vaporization of water is a multiplier of the terms of the equation for conservation of mass (Favre et al., 1976, 312).

The general equations of fluid mechanics, the Navier-Stokes equations (see II.2.1 and II.4), are thus appropriate for the atmospheric-hydrospheric system with state equations for humid air and saltwater. Since the number of equations is equal to the number of unknowns, these equations are closed and satisfy the general conditions for physicomathematical determinism. Nevertheless, in order to deal with the phenomena particular to the system, it is convenient to define and even to complete the equations.

With regard to electromagnetic radiation, the significant effects of absorption and emission are represented in the equation for internal energy by the divergence of the corresponding flux. This flux depends upon complex phenomena in the environment—the absorption and reflection of radiation by water vapor and carbon dioxide in the air, by the ground, and by the clouds; these are described by integrodifferential equations (Favre et al., 1976, 332–35). The global coefficients of absorption or emission of mass that result from them are proportional to the concentrations. This process is complicated but is still determined by the laws of physics.

Within the troposphere and at the surface of the oceans, where phase changes of water occur (condensation or evaporation, freezing or melting, phenomena with important effects on the reciprocal transformations of latent heat into sensible heat), complementary equations are necessary in order to determine the temperature and the mass that undergoes these transformations. These equations are based on equations for the conservation of the mass of each constituent and on their state equations—equations for density, pressure, and temperature in view of the physical constants (the heat of vaporization and melting and the critical values for the physical states of water, phenomena from the domain of thermodynamic determinism; see Mieghem and Dufour, 1948). Delays in phase changes may occur, but these changes are nearly null in the stratosphere and, because the particles of condensation are abundant, have practically no effect in the troposphere.

To sum up: the inanimate physical system of the atmosphere and hydrosphere satisfies the general conditions for physicomathematical determinism because of the applicability of the general equations of fluid mechanics and the complementary equations prescribed by the laws of the physics of radiation and the laws of the thermodynamics of changes of state, even if the complexity of certain phenomena obliges us to simplify the explanations for several of these equations. These general conditions are necessary because the behavior of water and air may not contravene the natural laws expressed by mechanics and physics.

III.2.1.2. The special conditions due to the particular circumstances of the fluid system under consideration must be analyzed in order to identify and describe in mathematical language the factors having the most significant effect upon the atmosphere and hydrosphere.

III.2.1.2.A. Boundary conditions are constantly imposed at the frontiers of the domain of the system, that is, at the level of the aforementioned conventionally adopted surface and at the level of the surface of the terrestial crust (the bottom of the ocean and the surface of the continents). This system is not isolated, and boundary conditions may be contingent without any incompatibility with determinism.

The conventional surface is permeable by matter, heat, and radiation. It is thus necessary to take possible vertical fluxes into account. With regard to transport of matter and sensible and latent heat, the mean fluxes are negligible because the mass of air exterior to that surface does not amount to more than 1 percent, because turbulence is nearly nonexistent at that altitude, and because the winds are generally horizontal. On the other hand, the vertical fluxes of radiant energy are significant. Incident solar radiation, averaged over the total surface of the planet outside the atmosphere, corresponds to a power of 340 w./m.^{2} per unit of terrestrial surface. As for the energy radiated by Earth, we know that it balances global solar radiation. However, there exists a significant excess of solar energy at lower latitudes and a notable deficit of solar energy at higher latitudes, which is compensated by the turbulent transport of energy along the meridians.

Cosmic rays also penetrate the atmosphere. They are composed of particles, electrons, hyperons, mesons, neutrons, nucleons, and photons, which travel at great speeds. They come in part from the sun and in part from interstellar and intergalactic space. Even though their global energy is relatively small, they seem to produce significant biological effects, perhaps by inducing genetic mutations.

The surface of the terrestrial crust is the site of a flux of geothermal energy on the order of 0.05 w./m.^{2}, which is very weak compared with the flux of solar energy, except at certain hot spots. The transport of solid mass is accomplished mostly by the erosion of the surface of the continents and the deposit of that material in seabeds as sedimentation. That transport is accomplished by rivers, glaciers, and coastal erosion. These phenomena are part of the geochemical cycle and ensure that the composition of seawater remains nearly constant. This cycle is characterized by *regularization* (Bureau des Longitudes, 1984).

The ground surface that projects from the seas is the site of exchanges with the atmosphere that depend upon the nature of the surface — its topography and vegetal covering: glaciers, deserts, steppes, mountains, savannas, cultivated land, zones of thick vegetation, and forests.[2] Fluxes of matter concern water and, to a lesser extent, oxygen and carbon dioxide from photosynthesis and combustion. Water transport forms a cycle: evaporation, melting or condensation, solidification or precipitation. On dry land, at sea level, annual precipitation is on the order of 670 mm., annual evaporation is 420 mm., and flows and subterranean infiltrations are 250 mm.[3] Of the water that arrives at the land surface, 50 percent is running water, of which 10 percent infiltrates the soil, and the other 50 percent evaporates — 13 percent directly from the soil, the other 37 percent from vegetation. Fluxes of sensible and latent heat across the surface of the land are equally important, with the flux of latent heat predominating in regions covered with vegetation and the flux of sensible heat predominating in deserts and over urban areas. Kinetic and convective turbulence intensifies these exchanges of matter, of sensible heat, and of latent heat by a regulatory action and also by limiting extremes of temperatures and concentrations of water vapor and carbon dioxide.

The very complex properties of the vegetal covering belong to the domain of biology; but the question that here concerns us is, What are the mechanical, physical, and chemical effects that this thin layer exerts upon the atmosphere? These effects may be formulated by modeling at different scales. At the micrometeorological scale, that of humans, models have been established to study the atmospheric-vegetal-ground exchanges for the mechanical and agricultural sciences. At the meteorological scale (say, greater than 1 km.), less detailed models are required; this exchange can be satisfactorily described by average coefficients of friction, of the transfer of sensible heat, of evaporation, of geometric shapes, and of the distribution and composition of the ground and its vegetal covering.

Chemical phenomena — particularly those accompanying the turbulent combustion of undergrowth, forests, and fossil fuels — are also representable by the Navier-Stokes equations of fluid mechanics, along with additional equations that take chemical reactions and their thermal effects into account (Libby and Williams, 1980).

Humans act upon the vegetal covering by deforestation and agriculture, by setting and extinguishing brush and forest fires, and by inducing desertification. Such action is contingent and even exhibits free choice, while remaining compatible with the determinism (as defined in I.5.5 and II.2.1) exhibited by the atmospheric-hydrospheric system. This fact lends humans both power and responsibility, because they can improve or destroy the natural medium of life.

III.2.1.2.B. Forces acting upon the system from a distance include that of Earth's gravity, which is nearly constant, and that of the moon's gravity, which is completely determined. The latter is very small with respect to terrestrial gravity, $10^{-7}\ \rho g$, but it causes the well-known daily and twice daily oceanic tides.

III.2.1.2.C. The state of the atmospheric-hydrospheric system at a given moment is, itself, a physical fact. The inadequacies of human observational techniques do not permit a description of that state in all its details but only at either the planetary, meteorological-oceanological scale or the micrometeorological scale. This does not affect the future behavior of the system or its deterministic character (see I.5.6 and II.2.1).

III.2.1.3. To summarize: the fluid system constituted by the atmosphere and hydrosphere is perpetually driven by turbulent flows and satisfies the general conditions for physicomathematical determinism. The complexity of certain phenomena like radiation and changes in the phases of water, however, obliges us here to shorten the description of the additional equations. These equations are necessary according to the general laws of physics.

The system also satisfies the particular conditions regarding forces acting from a distance and regarding its state at a given moment. With regard to boundary conditions that reflect the frontiers of the exterior medium, they may easily be formulated for a surface adopted by convention at an arbitrary altitude and for the depths of the oceans. The surface of the ground is heterogeneous and more complex, especially when covered by vegetation. Its formulation can be modeled at different scales to express its mean physical effects upon the atmosphere. Such modeling is standard at the scale of meteorological observations.

One may consider the behavior of the generally turbulent, inanimate, atmospheric-hydrospheric system to be physically and chemically determined. Physical-chemical determinism is compatible with contingency and even with freedom of human action (see I.5.5 and II.2.1).

III.2.2. The atmospheric-hydrospheric system may also be considered to be regulatory and optimizing.

III.2.2.1. Earth annually receives and emits equal quantities of radiated energy. This results from global regulatory mechanisms in which the atmospheric-hydrospheric system plays an important role, in particular by the vertical and horizontal turbulent transport of mass and energy.

The scientific study of this system is based on statistical equations for fluids with variable densities (see II.4.1) and on additional equations that take the special phenomena of radiation and change of phase of water into account (see III.2.1). Let us recall that these statistical equations explicitly describe the divergence of the mean fluxes of ρw transported by three mechanisms: the mean mass-weighted motion, $\bar{\rho}\tilde{w}\tilde{u}_k$, turbulent fluctuations, $\overline{\rho w' u'_k}$, and molecular agitation, $\overline{H}_k$. The ρws represent, by unit volume:

mass, ρ, momentum, ρu_i, internal energy of sensible and latent heat, ρe, and the mass of the water vapor in the atmosphere or of the salinity in the oceans, ρc. The three mechanisms for the transport of ρw tend to homogenize momentum, internal energy, and concentrations of water in the atmosphere and salt in the oceans, thus limiting the extremes of speed, temperature, humidity in the air, and salinity in the oceans. These are deterministic regulatory mechanisms that optimize the terrestrial climate.

The most significant transports of energy and mass are produced by mean movement and turbulent diffusion. The effects of molecular agitation are negligible except where there is contact with the frontiers of the domain, because the Reynolds parameter reaches very high figures in the atmosphere and oceans. The vortices that carry energy are very large in comparison with the vortices within which molecular viscosity manifests itself.

The quantities of sensible and latent heat that are transported are large. They are composed of net excess radiation absorbed at latitudes below 35° transported to the area of net heat deficit at latitudes above 35°. Now, the part of that quantity of mean and turbulent kinetic energy that accomplishes this transport and that is dissipated by friction as heat is relatively small when the Reynolds parameter is high. This finding corresponds to an optimization in the sense of a minimal cause producing a maximal effect.

III.2.2.2. The particular conditions of this system (see III.2.1) and the observations at both micrometeorological and meteorological-oceanological scales permit simplifications of the general statistical equations.

The atmosphere and the oceans constitute thin layers with respect to their horizontal dimensions. The motions observed at a meteorological-oceanological scale are practically horizontal. On the contrary, however, the effects at the surfaces of the continents are significant, and the vertical components of the gradients are much greater than the horizontal components of pressure, velocity, temperature, concentration, and density in the case of the atmosphere.[4]

III.2.2.3. With regard to the vertical component, gravity and the gradient of mean pressure nearly compensate for each other. The vertical distribution of mean pressure is thus the same as it would be were velocity to vanish — which is to say, there is a situation of hydrostatic equilibrium.

In the oceans, since the density is constant up to nearly a few parts per thousand, the equations easily describe the vertical distribution of mean pressure, according to the linear law just mentioned.

In the atmosphere, density and pressure decrease markedly with altitude; they are related to temperature by a state equation. Temperature is given by the equation for internal energy, taking possible phase changes into account. A particle of air, ascending or descending rapidly, hardly exchanges any thermal energy with the environment; the process is nearly *adiabatic.* Air cools when it dilates and warms when it compresses at a rate of $-g/c_p$, or 1°C per 100 m. of altitude if there is no condensation or evaporation of water, and at a rate of 0.65°C per 100 m. if the air is saturated and phase changes occur.

If the vertical distribution of mean temperature in the atmosphere exhibited a decrease greater than 1°C per 100 m., the ascending particles would be warmer and less dense than the surrounding air, and by Archimedes' force, they would continue their movement and there would be instability. If, on the contrary, the vertical decrease were less than 0.65°C, the ascending particles would be colder and denser than the surrounding air, they would tend to descend to their original level, and there would be a stable equilibrium. When the vertical gradient of atmospheric temperature is between -1°C and -0.65°C, there is instability if the air is saturated, because condensation releases the latent heat of the water vapor, which forces the air to rise. It is for that reason that cumulonimbus clouds reach altitudes of above 10 km. Up to the height of the tropopause, the atmosphere harbors turbulent convection, which produces significant vertical transport of thermal energy according to a regulatory mechanism for temperature.

For regions where cloud formations are numerous, one defines standard atmosphere as that characterized by a mean temperature of -0.65°C and, at sea level, by a pressure of 1,013.25 millibars and a temperature of 15°C. Then the equations permit a formulation of a reference density and pressure for the troposphere. For the lower stratosphere, the formulas suppose a zero temperature gradient. The differences between the mean temperature gradient of the atmosphere and standard values permit a determination of zones of instability and of the heights of clouds. Clouds have a regulatory effect upon temperature, because they absorb some incident solar radiation and radiation emitted by Earth's surface.

III.2.2.4. With regard to motions at the meteorological-oceanic scale, velocities are represented by winds and currents, **V**, which are horizontal, and the governing forces are the horizontal gradient of pressure, **G**, and the horizontal components of the Coriolis force, **C** (Favre et al., 1976, pt. 6; Phillips, 1977). The centrifugal inertial force, $\rho V^2/R$, intervenes when the radius of curvature, R, of the horizontal trajectories is relatively small. It is often disregarded up to a first approximation. The total constraints of viscous friction and turbulent tension are negligible outside the boundary layers. The boundary layers, within which kinetic turbulence develops, have a thickness of about 1,000 m. in the atmosphere above the ground and the surface of the oceans. In the oceans they are thinner and develop near the surface in the currents of the Ekman drift driven by the wind (Bureau des Longitudes, 1984).

The horizontal component of the Coriolis force is perpendicular to the velocity, which lies to its left in the Northern Hemisphere and to its right in the Southern Hemisphere. Its modulum is $2\omega\bar{\rho}V \sin \varphi$, where ω is the velocity of the rotation of Earth, V is the modulum of the mean velocity of the fluid, and φ is the latitude. The horizontal component of the Coriolis force is null or negligible at the equator and increases with latitude.

The ratio of the forces of inertia to the Coriolis force is represented by the Rossby parameter of similitude, $V/2\omega D$, in which D is the characteristic length of the flow. The Coriolis force is thus preponderant when this parameter is small with respect to 1. This is the case in the atmosphere and in the oceans outside of the equatorial zone for large-scale motions, D. For meteorological perturbations when the characteristic velocities are 20 m./sec. and the characteristic lengths are 1,000 km., this parameter is 0.14. In ocean currents like the Gulf Stream, which has velocities of 2 m./sec. and undulations of 100 km. in length, this parameter is still 0.14.

Up to a first approximation, and for large-scale motions, we may suppose that the Coriolis force and the gradient of pressure balance each other except in the equatorial zone and in boundary layers. The Buys-Ballot law then holds that the "geostrophic wind or water current," **V**, is tangent to the contours of equal pressure, the isobars, leaving the lower pressures to the left in the Northern Hemisphere and to the right in the Southern Hemisphere. Their modula are proportional to the gradient of pressure and inversely proportional to the density and sine of the latitude, $G/2\bar{\rho}\omega \sin \varphi$.

In the Northern Hemisphere, the wind or the current turns clockwise around a center of high pressure and counterclockwise around a center of low pressure. In the Southern Hemisphere, the rotation is reversed. Subtropical cyclones, hurricanes, and typhoons are low-pressure vortices around which the winds blow counterclockwise in the Northern Hemisphere and clockwise in the Southern Hemisphere. Winds are extremely strong in the lower latitudes, where cyclones are formed, and diminish in strength by the action of the Coriolis force as they move into higher latitudes. At a given latitude, momentum, $\overline{\rho \mathbf{V}}$, depends exclusively upon the gradient of the pressure. In the case where it varies slightly, since $\bar{\rho}$ decreases rapidly, winds increase with altitude, $\overline{\rho \mathbf{V}}$ being nearly constant.

In the equatorial zone outside the boundary layers the governing forces are pressure gradient and centrifugal force. The cyclostrophic velocity, or current, is still tangent to the isobars, and its modula are $(GR/\bar{\rho})^{-1/2}$. The sense of rotation of the vortices is not fixed within latitudes between $+2°$ and $-2°$.

In the atmosphere, the geostrophic wind corresponds quite well to the real wind outside the boundary layers and the equatorial zone. In the ocean, the geostrophic current corresponds rather well to the mean current, except near the surface and the seabed. However, the wind exerts a constraint upon the surface of the water, which leads to a "drift" current. And in any event, the effects of the tides, the boundary conditions imposed by the ocean bed, and the shores modify the currents.

The distribution of the winds and currents within the boundary layers depends not only upon the gradients of pressure and the Coriolis force but also upon the divergence of total constraints of molecular friction and, above all, upon turbulence stresses (see II.4.2.2). In the atmosphere, the divergence of the resulting constraints, which act in the opposite sense to velocity, produces a decrease in velocity and deflects it toward the lower pressure, according to Ekman's spiral. They contribute to the decay of low-pressure zones. At the surface of the ocean when there is no pressure gradient, a uniform wind produces a drift current on the order of several percent of its velocity. Because of the effects of the Coriolis force, this current flows at an angle of 45° with respect to the direction of the wind, to the right in the Northern Hemisphere and to the left in the Southern Hemisphere. As the depth increases, the drift current decreases; it vanishes at a depth of 100 m., according to Ekman's spiral (Bureau des Longitudes, 1984). The effects of the wind combine with those of the geostrophic current when the pressure gradient is not null.

The masses of a single fluid, whether air or water, coming from different sources with distinct densities, ρ_1 and ρ_2, and velocities, V_1 and V_2, are separated by discontinuity surfaces, which persist for a long time. The pressures are equal at each side of the discontinuity surface. If that surface moves with a uniform velocity, the dominent forces (outside the equatorial zone) are the horizontal components of the Coriolis force and the weight of the mass of each fluid. The Margules formula gives the slope of the discontinuity surface.[5]

In the atmosphere, caps of cold air cover the polar regions at low altitudes, with layers of relatively warm air above them that extend to the equator. These caps of polar air are separated from tropical air by two discontinuity surfaces, the polar boreal and the austral front, whose slope is about 1 percent. When these fronts undergo perturbations whose wave lengths are longer than 1,000–1,500 km. and whose amplitude exceeds 200 km., they become unstable (Helmholtz). The perturbations are propagated from west to east and are transformed into large vortices; these are the low-pressure meteorological perturbations well known in the middle latitudes. The winds in these vortices blow counterclockwise in the Northern Hemisphere and clockwise in the Southern Hemisphere.

In the oceans, water from different sources may also have different masses and different velocities. These waters are also separated by discontinuity surfaces whose slope corresponds to the above formula. This is the case for a current of water from a strait into an ocean, such as Mediterranean water flowing into the Atlantic Ocean along the Portuguese coast. It drifts to the right, driven by the Coriolis force (Lacombe, 1965).

III.2.2.5. To study the behavior of the atmospheric-hydrospheric system, we must describe however briefly its structures, its general circulation, and its average effects at the meteorologic-oceanological scale.

It is customary to separate atmospheric currents into averages taken over parallel latitudinal zones, in summer and winter, and to consider deviations between zone averages and real atmospheric winds measured at a meteorological scale.

At the zonal scale and in the lower latitudes, the predominant eastern winds have a force of between 0 and 10 m./sec. at all altitudes. At the intermediate latitudes there are stronger western winds. Near the poles, there are small areas of eastern winds. Western winds reach a maximum speed of 45 m./sec. in the winter at a latitude of 30° and an altitude of 13 km. These winds are known as *jet streams*. Along the meridians, the atmospheric dynamic is characterized by three cells in each hemisphere. Between the tropical cells, intense updrafts of turbulent convection carry with them large quantities of heat and humidity. Between these cells and those in the intermediate latitudes are downdrafts. Finally, in the upper latitudes are the polar cells (Bureau des Longitudes, 1984).

At the meteorological scale, the thermal contrast between the oceans and the continents has a very strong effect upon winds in the lower layers of the atmosphere. The continents are warmer than the oceans in the summer and colder in the winter, and this modifies the distribution of atmospheric pressure and, thus, winds — mainly in the Northern Hemisphere, where the continents occupy relatively more of the surface above 40° latitude. Figure 10 depicts the mean characteristics observed in the lower atmosphere during July and January: the high and low pressure centers, isobaric curves, wind pressures, and fronts, which are discontinuity surfaces between different masses of air generating meteorological perturbations.

Nearly permanent air pressure centers are situated above the oceans: these include, at the subtropical latitudes, the five anticyclones of the northern Pacific Ocean, the southern Pacific Ocean, the northern Atlantic Ocean,

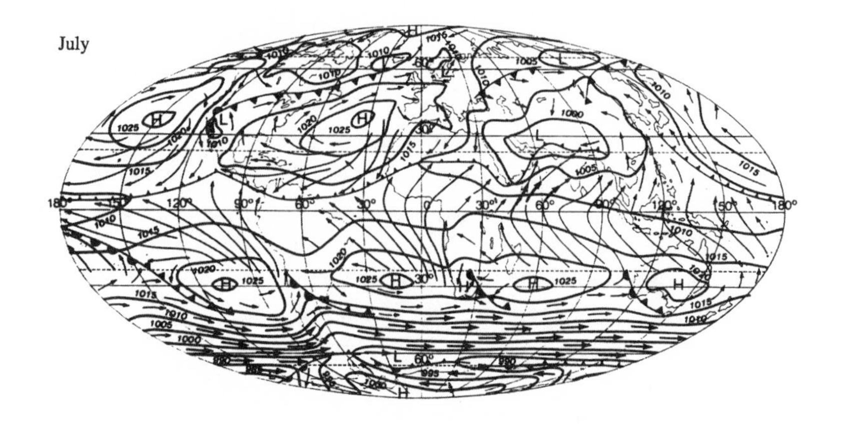

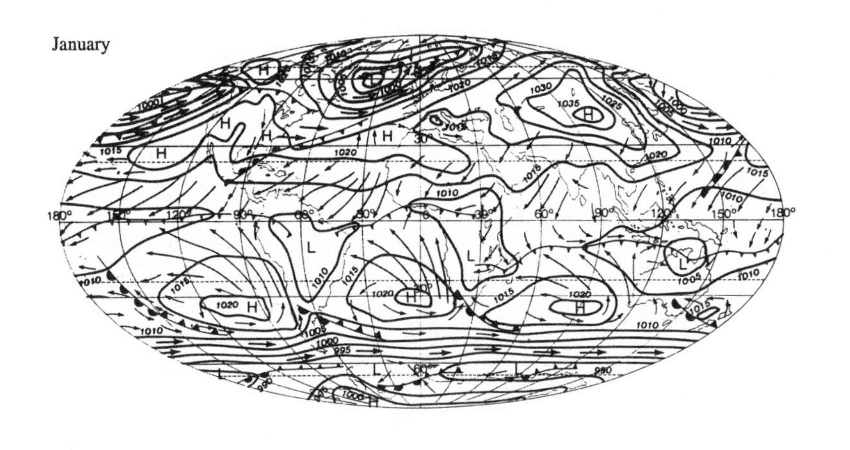

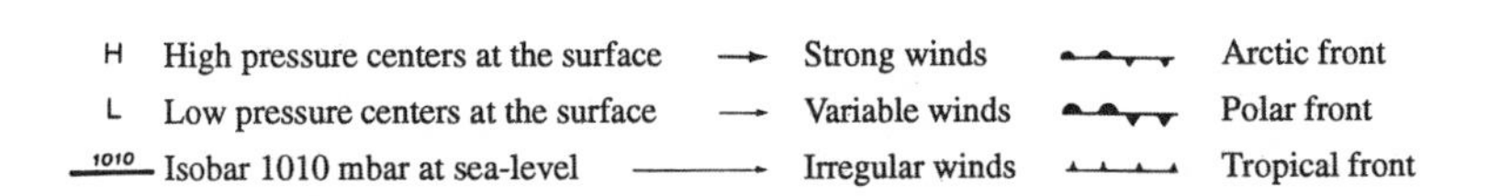

Figure 10. The characteristics of the lower layer of the atmosphere, July and January. From Bureau des Longitudes, 1984.

the southern Atlantic Ocean, and the Indian Ocean; and in the intermediate latitudes, the belt of low pressure in the Southern Hemisphere and the two low-pressure centers in the Northern Hemisphere in the Pacific and the Atlantic. Seasonal pressure centers form over the continents. They are anti-cyclones in the winter and low-pressure centers in the summer, principally over Siberia but also over Australia and Canada because of their variations in temperature.

The Buys-Ballot law permits us to deduce the distribution of winds from this distribution of pressure. Winds are tangent to isobars at an altitude of 1,000 m. and deviate toward low-pressure zones at altitudes of several tens of meters. In the Northern Hemisphere, winds blow clockwise around the anticyclones and counterclockwise around low-pressure centers. These directions are reversed in the Southern Hemisphere. This explains the northeast and the southeast tradewinds in belts north and south of the equatorial doldrums, where the dominant winds blow from the west in the intermediate latitudes and from the east near the poles. But over the continents, winter anticyclones are replaced by low-pressure centers in the summer, the source of monsoons. In southern Asia in winter, cold, dry monsoons predominate, blowing from the northeast; in summer, warm, moist monsoons blow from the southwest.

Cyclones form in subtropical oceanic regions. Their paths wind around the oceanic anticyclones to the west, clockwise in the Northern Hemisphere and counterclockwise in the Southern Hemisphere. The two polar fronts, boral and austral, are traversed by low-pressure perturbations from west to east.

The general oceanic flow may be represented schematically by distinguishing surface currents from deeper currents. Surface currents correspond, up to a first approximation, to dominant winds. They flow clockwise around the anticyclones of the northern Pacific and Atlantic Oceans and counterclockwise around those of the southern Pacific and Atlantic Oceans and the Indian Ocean. A constant westerly flow exists in the Southern Hemisphere in the intermediate latitudes. Near the continents, the currents follow the shoreline, thus creating coastal currents to the west and east of these oceans. Alternate seasonal currents are created by monsoons, especially in the Indian Ocean. These surface currents penetrate to a depth of 100 m., and their velocity is greater in the west (where they achieve speeds of a meter a second) than in the east.

Deeper water currents are related to the density gradients due to surface thermal fluxes. Winter cooling of surfaces, in especially the Atlantic Ocean and the Mediterranean Sea, produces a strong convection current over small areas (3 percent), whose vertical velocity may attain 10 cm./sec. Currents are created from these "cold springs," mostly along the meridians, at depths of several thousands of meters, with velocities of around 1 cm./sec.

The energy of solar radiation, which is maximal near the equator, produces a warming in subtropical zones, which together with the seasonal thermal contrast between the oceans and the continents drives the sea currents.

The world's oceans constitute an immense heat reservoir and thus contribute to reducing the daily and annual differences of temperatures on the planet. The continents, where solar energy penetrates less deeply, undergo greater variations of surface temperature.

In the atmosphere, vertical fluxes of sensible and latent heat are carried by turbulent kinetic diffusion from the surface throughout the boundary layers. Were there no turbulence, the effects of molecular diffusion alone would prove too weak to carry the heat, because the thermal conductivity of air is low. The nonturbulent boundary layers have thicknesses of several meters, while the turbulent layers have thicknesses of several hundreds of meters. Fluxes of sensible and latent heat in turbulent flows are greater than those in exclusively molecular diffusion by several orders of magnitude. Fluxes of turbulent convection, which may attain 10 km. in altitude up to the tropopause, are very significant. Were there no turbulent winds and currents, the ground temperature might easily exceed the boiling point of water, 100°C, at normal atmospheric pressure. A dense layer of carbonic gas several meters thick would cover Earth's surface, as mentioned earlier.

In the oceans, the vertical flux of sensible heat is triggered by kinetic turbulence in the boundary layers, down to a depth of about a hundred meters, and the flux of convection penetrates several kilometers in depth, thus circulating heat, oxygen, and marine salts.

Turbulent winds and currents carry enormous quantities of sensible and latent heat horizontally from the equatorial zones toward the polar zones, for whose heat deficiencies they compensate in proportions estimated at 60 percent and 40 percent. They also carry large quantities of oceanic thermal energy toward the continents. In the course of their winding around high- and low-pressure centers, winds and currents in the lower latitudes receive sensible and latent heat from the air and sensible heat from the water, which they carry great distances to the intermediate latitudes. The northern and southern polar fronts are also very active sites for the movement of warm, humid air from the tropics toward the poles and cold, dry air from the polar regions toward the intermediate latitudes.

Ocean currents that arise in the tropics and wind westward around oceanic low-pressure centers—like the Kuroshivo current in the northern Pacific and the Gulf Stream in the northern Atlantic—carry heat and thus improve the climate along the east coasts in the lower latitudes. They then cross the oceans and moderate the climate along the west coasts in the intermediate latitudes. Westerly winds from the oceans that dominate in the intermediate latitudes thus moderate the climate of the western regions of the continents.

As for subtropical cyclones that form over the oceans, they also carry thermal energy vertically and horizontally. The winds are very strong at the lower latitudes and may produce natural disasters, but when they reach the intermediate latitudes, after winding around the oceanic anticyclones from the west, their strength decreases because of the effects of the Coriolis force. The slowing down occurs optimally, without notably diminishing the winds' energy, because since the Coriolis force is orthogonal to the speed, it performs no work. Since the kinetic momentum is conserved, the diameter of the cyclone increases when the velocity of the winds decreases. Subtropical cyclones in the intermediate latitudes to the east of the oceans cannot be distinguished by the strength of their winds from the depressions of polar fronts.

Evaporation and condensation absorb or release great quantities of latent heat, which is maximal for water. These phenomena help reduce temperature extremes. This is also true for the melting and freezing of water. The cloudy layers thus formed in the atmosphere absorb incident solar radiation and the radiation reflected back by Earth; they play a role, together with water vapor, in the greenhouse effect, limiting extremes of temperature. Condensation of water vapor and cloud formation could not develop were it not for turbulence, which diffuses the particles about which condensation forms throughout the troposphere. The regulatory effects of homogenization and the limitation of extremes of temperature and internal energy on a planetary scale lead to a limiting of the pressure gradient and, thus, a limiting of winds and water currents. The latter weaken progressively in the higher latitudes because of the Coriolis force and in the boundary layers of the surface because of viscous friction and turbulence. Turbulence also reduces the velocity of flows in streams and rivers.

The atmospheric-hydrospheric system is the scene of complex phenomena with turbulent flows, among which are numerous nonlinear interactions. Its mechanisms are represented by closed equations and are thus physically and chemically determined. The average effects of these mechanisms contribute to the regulation and equilibrium of terrestrial energy and to the homogenization of the composition of the air and water, to the following limitations: extremes of temperature; concentrations of carbon dioxide, water vapor, ocean salinity, and pollutants due to human action;[6] and velocities of winds and water currents. Without excluding strong local and temporary variations in temperature, humidity, wind, water current, and precipitation, the average effects of the functioning of the atmosphere and hydrosphere correspond to a global regulation and optimization of the possible states of terrestrial climate.

III.3. Conclusion

Earth is the only planet in the solar system possessing an atmosphere and a hydrosphere, favorable media for life.

Oxygen is abundant in these media. In the atmosphere, it is mixed with nitrogen; in the hydrosphere, it is in solution. Water is also abundant in its three phases — liquid, vapor, and solid. The principal constituents of the biological cycles — carbon, hydrogen, sodium, chlorine, calcium, magnesium, potassium, and so on — are also present. Oxygen, hydrogen, and carbon have maximal properties: heat for vaporization and formation, the capacity to dissolve various substances, surface tension (for water), separation between water and air, chemical neutralization (for carbonic acid), and the number and diversity of their chemical compounds.

Energy is supplied mainly by radiation, which is inversely proportional to the square of the distance from the sun. Earth orbits the sun at an optimal distance, which varies only slightly, because its orbit is not very eccentric. Because of that slight variation, however, it receives some additional energy in January. Its rotation about an axis that is nearly perpendicular to the ecliptic distributes the radiation along the latitudes every twenty-four hours. The slope of that axis gives rise to seasons; and all zones, even polar zones, receive a minimum quantity of energy annually.

Part of the solar radiation arrives as light. Light is only minimally absorbed by the atmosphere, but it is absorbed by the vegetal process of photosynthesis, thus forming organic molecules in green plants by the reduction of the pair, carbon dioxide and water, with a release of oxygen. On the contrary, the narrow band of ultraviolet radiation, which, as already mentioned, is inimical to life, is absorbed by the atmospheric ozone.

Finally, the partial absorption of infrared radiation by water vapor, carbon dioxide, and clouds regularizes temperatures in the atmosphere. Earth annually receives and radiates equal quantities of energy, so it is in a global energy equilibrium.

The atmospheric-hydrospheric system is very complex, with flows that are generally turbulent. Its behavior is deterministic and conforms to the laws and fundamental principles of classic mechanics, physics, and chemistry, formulated in equations whose number equals that of the unknowns. These laws and principles are among the most completely verified elements of scientific knowledge. The phenomenon of turbulence, despite its complexity, is governed not by chance or disorder but by an order that changes as the scales and structures change. This determinism implies the necessity of its conformity with these laws and principles. It also takes into account the particular circumstances of Earth's state at a given moment: the effects of gravity and lunar and solar attraction and the boundary conditions at an arbitrarily chosen surface (for example, the surface at 30 km. altitude, the ground surface, and the surfaces of the ocean depths). These boundary conditions may be contingent because of the variation of solar and cosmic radiations and especially because of the variations in the vegetal covering of the continents.

The apparent freedom of human beings may be recognized in their actions upon the environment through deforestation, agriculture, the burning of vegetation and fossil fuels, the production of desertification, and the pollution of the environment.

The transport of mass and energy by the movement of the atmosphere and oceans has mean global regulating effects, in which turbulence plays a role in diffusing mass, energy, and momentum (just as molecular motion does, but more intensely). These effects tend to homogenize physical properties, to limit extremes of temperature, to limit concentrations of oxygen, carbon dioxide, water vapor, and salinity, and to limit the force of winds and water currents.

Vertical transport is governed principally by the diffusion produced by kinetic and convective turbulence in the troposphere. The general circulation of the atmosphere and oceans is dominated by the gradients of pressure and by the Coriolis force, which coils winds and water currents around high-pressure centers, transports enormous quantities of thermal energy from the equator toward the poles, and contributes to planetary energy equilibrium.

The phenomena just discussed—whose enumeration has not been exhaustive—interact to produce global regularization and optimization of those physical and chemical states that are possible and compatible in the nonliving terrestrial material medium. The criterion of optimization is evident in the coordination between the states of the atmosphere and hydrosphere, on the one hand, and the conditions necessary for the support of terrestrial life, on the other. This coordination cannot be explained by an effect of chance, which would imply the independence of the numerous causal chains encountered in the study of the atmospheric-hydrospheric system, or by an effect of disorder, which would imply the independence of numerous parts and properties of the system. In fact, the occurrence of such coincidences would have a probability equal to the product of the probabilities of each of its independent elements. Because of the great number of elements, that probability would be extremely low, practically null, which shows that these numerous chains and properties are not independent.

The fact that the convergence of so many factors has occurred and has continued for so long implies connections between the behavior of nonliving matter and that of living organisms. These connections have a teleonomic character, without which life on Earth could not be maintained, could not develop, and could not evolve.

Such connections could be explained in part by the adaptation of animate organisms to the physical-chemical conditions in the atmospheric-hydrospheric medium as a result of natural selection and evolution, as has been documented for variations in terrestrial climate over the ages. Biologists have discovered regulatory mechanisms and teleonomic behavior in animate organisms (see chapter V). But this adaptation, which includes bifurcations, seems possible only within certain limiting conditions. Deviations from the "normal" climate must be small, and variations must be quite slow. Thus the variation in mean global temperature since the last ice age has been on the order of only a few degrees centigrade. Certain thresholds — the boiling point of water, minimal concentrations of oxygen in the air and water, a minimum of light, a maximal ultraviolet intensity, the maximal salinity of water, a lack or a superabundance of certain elements that are major constituents of living matter, the maximal effects of winds and water currents, which produce forces proportional to the squares of their velocities — may not be crossed. According to the most recent observations, no other planet in the solar system — not even Mars, whose fluid and climatic medium is closest to Earth's — harbors life. The adaptation of organisms to the exterior medium surely exists, but it does not appear sufficient to explain the coordination that has been demonstrated. We are also led to seek such connections through the behavior of the nonliving medium, which conforms to the laws of physics and chemistry in that ecosystem.

The mechanisms of the atmospheric-hydrospheric system are deterministic because they satisfy the general laws of physics as applied to their particular circumstances. Among the latter are the effects of gravity, of lunar attraction, and of radiation arriving from the solar system, and these are governed by the laws of physics (see chapter IV). Cosmic radiation, coming mostly from more distance sources, is less well known, but its energy effects on the atmosphere and the ocean seem weak.

The behavior of the vegetal covering obeys the laws of physics, chemistry, and biology. It plays a regulating role in the coordination of terrestrial nature but one that does not, in itself, suffice to explain that coordination.

The behavior of the atmospheric-hydrospheric system is consistent with deterministic equations required by the classic laws of mechanics, physics, and chemistry. Principles from classic mechanics (conservation of matter, energy, and variation of momentum; see chapter IV) and the more synthetic principle of least action from analytical mechanics and physics (in particular, from mechanics and electromagnetic radiation) express in different mathematical terms the same rules of optimal and compatible variations.

Theoretical physics shows that these principles can be interpreted both as *causalist*, because of their formulation as differential equations, and as *teleonomic*, by the calculus of variations (see chapter IV). The theory of dynamic systems tends to confirm the complementary relation between efficient causes and teleonomy (see I.3). Thus diverse methods concur in discovering a convergence between the behavior of the nonliving terrestrial medium and the behavior of living organisms toward a global, teleonomic arrangement.

The conditions for coordination are not always satisfied, because natural, local catastrophes sometimes destroy animate organisms, but the criterion is an optimization of the compatible, possible states that permit the global support, development, and evolution of life upon Earth.

One might imagine life existing in other places in the universe, but the probability for that to be the case is at most equal to the sum of the probabilities of the existence of other stellar systems that possess media and climates favoring the establishment and support of life. This probability increases with the number of stars considered, which may be quite large. But as the distances to these stars increase with their number, so the possibilities of observation, communication, and access diminish. Furthermore, it is not certain that the forms of life that might exist in those stellar systems would sufficiently resemble terrestrial life to permit identification of the two and communication between them. Our study has been limited to the observable terrestrial medium and to its relationship with life as we know it.

Notes

1. After the appearance of his books, *The Fitness of the Environment* (1913) and *The Order of Nature* (1917), Henderson delivered a lecture with an interdisciplinary perspective before the French Philosophical Society (Henderson, 1921).

2. As is customary, we suppose here that the mechanical effects of plant and animal life on the atmospheric-hydrospheric system are globally negligible. This is not always the case, however, because vegetation affects the carbon dioxide balance, which in turn influences the greenhouse effect.

3. Between the atmosphere and the ocean, precipitation is on the order of 870 mm. and evaporation 970 mm. For the continental-oceanic-atmospheric system, the annual balance of the global water cycle is zero. This is yet another regulatory mechanism.

4. In the interior of the system, the conventional surface separating the atmosphere and the oceans also harbors exchanges of mass and energy. These are the subjects of specialized studies, based upon the same principles (Favre and Hasselmann, 1978).

5. According to the Margules formula, it is $-2\omega \sin \varphi(\overline{\rho}_1 V_1 - \overline{\rho}_2 V_2)/g(\overline{\rho}_1 - \overline{\rho}_2)$.

6. See Taylor, 1921. The diffusion in a turbulent flow may, in a first approximation, be described as proportional to the distance downstream near the source, where the correlation along the paths remains strong, and proportional to the square root of the distance downstream far from the source, where the correlation vanishes (Hinze, 1975, 48–55). So turbulent flows in the atmosphere propagate pollution with a passive contaminant short distances from the source and then clean the atmospheric-oceanic system by increasing dilution at a long distance from the source.

IV

Physical Theories: Groups, the Calculus of Variations, Relativity, Statistical Mechanics, and Quantum Mechanics

There are many ways of using mathematics to obtain a better understanding of concrete reality. In some of them, mathematics is invoked only to provide an extrinsic tool. The scientist wants to arrive as quickly as possible at a numerical comparison of experiments and thus to assert his or her power over things. But it is not this aspect that interests us here.

In the best-developed areas of our science, mathematics often plays a more ambitious and necessary role. It becomes a way of thinking, in which reality is apprehended on the basis of mathematical concepts, and one does not claim to understand reality until one knows how to construct a consistent and effective mathematical model for the set of phenomena being studied. The great physical theories of our time—relativity, quantum mechanics—like classic mechanics before them, are composed of such models, and the development of new mathematical concepts has often been closely linked to what was required to understand reality.

It may appear that a physical theory has two distinct aspects, one being esoteric and untranslatable except in a sophisticated jargon and by means of formulas, the other being popular and expressible in our usual language of everyday experience. However, the reality of such a theory is that it is based on mathematical concepts. It is fortunate when these concepts are generally understood and the theory may be expressed in everyday language. But such a description is inevitably based on analogies, and they may be deceptive. However, as Poincaré points out, successful analogies stimulate our imagination and may lead to further scientific hypotheses.

Before discussing relativity or quantum mechanics, and to minimize the danger of misunderstanding, we must analyze the concepts—or rather the mathematical approaches—that play a universal role in physical theories.

IV.1. Examples of Mathematical Instruments of Thought: Groups and the Calculus of Variations

The idea of a *group*—a term he introduced—appears in the work of Galois around 1830, in a context far removed from physics—that of the solution of algebraic equations by radicals. Galois's groups have two roles: they are groups of permutations, that is, of transformations, of the roots of equations, subject to certain conditions; but they are also "abstract" groups, in which mathematical objects are both the elements and the way in which the groups are composed.

But throughout the nineteenth century, this concept of groups appeared in geometry and in mechanics, and it appeared toward the end of the century in the electromagnetic theory of Maxwell, essentially in the form of transformation groups. In fact, since antiquity, this understanding of groups has underlain certain areas of intellectual activity. In Greek geometry, culminating in Euclid, we see the axioms concerning "equality" of figures assert the following:

1. A figure is equal to itself.
2. If a figure is equal to a second figure, the latter figure is equal to the former.
3. If one figure is equal to a second figure, and that figure is equal to a third, the first figure is equal to the third.

The axioms of a ground underlie these assertions but in a form that is still awkward. It was only the introduction of the vocabulary of set theory that permitted the construction of a general definition of a *group.* If one has a well-determined set, E, one is led to study its transformations. A transformation, f, brings element a of E into correspondence with element b of E, and conversely. One thus obtains an inverse transformation. Two such transformations, f and g of E, may be naturally composed with each other associatively. The composition, $g \circ f$, is the transformation such that to each a in E there corresponds an element, $g[f(a)]$.

A set, G, of transformations of E will be a transformation group if

1. it always contains the composition of any two transformations of G,
2. it contains the identity transformation, that is, one that brings any element of E into correspondence with itself, and
3. it contains an inverse transformation for each transformation in G.

It is obvious that, exclusive of their order, these axioms correspond to those of the equality of figures in Euclidian geometry. G will be an (abstract) group if we start with a set that has an associative composition law for its members, a law for which, for each element of the set, there is a neutral or unit element (analogous to an identity transformation) and an inverse element. This is a brief description of a concept that dominates mathematics and that supplies one of the principal keys for understanding reality.

In these terms, which simplify our language, we can say that the Euclidian geometry of space depends entirely upon the *group of motions* of space, the group whose elements are obtained by composition of rotations and translations and whose transformations preserve distances. The translations themselves form a group, whose composition law, which is described by the addition of vectors, is commutative. Rotations about a point also form a group. A transformation group transforms certain elements or certain quantities, while others remain invariant, unchanged. This is the case for distance, for example, for the group of translations in Euclidian geometry—and this is important. Certain equations may remain invariant in their form while the quantities they involve change.

Since the discovery by Bernhard Riemann and Nikolai Lobatchevski of non-Euclidian geometries, they have been defined as a group, and this group has become an essential instrument for their analysis. This type of analysis has been extended to mechanics and electromagnetism.

Newtonian mechanics was first expressed in terms of mass, acceleration, and forces. This is particularly true of the Newtonian theory of gravity. But what was the group that left the equations of this mechanical theory invariant? And thus, what was the group that corresponded to this theory? It was recognized quite early that this was the group formed by the composition of spatial motions, of translations in time, and of uniform, rectilinear motions. One could say that this was the fundamental group of transformations of Newtonian space-time. This particular group, each of whose transformations is described by the values of ten parameters, is called the *Galilean group*, in homage to the scientist who first recognized the paramount importance of uniform, rectilinear motion.

The invariance of the equations of Newtonian dynamics under the Galilean group is the best way of expressing the principle of Galilean relativity, which asserts the impossibility of proving, by means of terrestrial mechanics, the absolute velocity of Earth—that is, its velocity with respect to a reference frame centered on the gravitational center of the solar system and fixed with respect to the stars. In its most primitive version, the Newtonian description of motion is apparently causalist, because the Newtonian concept of force plays the role of a cause for accelerations, that is, for the deviation of motions with respect to the uniform rectilinear motion associated with Galilean reference frames. From the point of view of classic mechanics, Galilean reference frames are transformed one into the other by the Galilean group. The result is an assertion of the principle of Galilean relativity. No purely mechanical experiment performed in a Galilean frame of reference can demonstrate the motion of that Galilean reference frame with respect to another Galilean reference frame.

Around 1865, James Maxwell succeeded in unifying electric and magnetic fields into a single entity, the *electromagnetic field*, governed by equations that bear his name. At the heart of this entity, the various forms of radiation recognized in his time or discovered in the next thirty years are united: light, of course, but also x rays, γ rays, and radio waves. This was the first great unification of domains in physics. The synthesis of electricity and magnetism showed the importance of a constant, c, which had the dimensions of a velocity and which theory identified with the velocity of the propagation of electromagnetic waves in a vacuum and, thus, with the velocity of light. Maxwell's equations led to a wave equation analogous to that of fluid mechanics, where c plays the same role as, for example, the speed of sound.

Around 1900, several theoreticians, among them Hendrik Lorentz and Poincaré, effectively determined an invariance group for the wave equation or, under suitable hypotheses, for Maxwell's equations, a group that seems profoundly different from the Galilean group. It was a first approach to what is now called the *Poincaré group*, upon which the theory of relativity is based. In the next section, we describe this theory using the physical concepts of that period and then return to the importance of the Poincaré group.

Another universal instrument is the *calculus of variations*. The prototype of this instrument appears early in elementary geometry when one proves that a straight line is the shortest path between one point and another. The first appearance of the calculus of variations in science was in Fermat's principle regarding the propagation of light. The index of refraction at a point in a transparent medium is actually the ratio of the velocity of light in a vacuum, c, to its velocity in the medium. Fermat asserted that the path of a ray of light, which might be curved by variations in the index of refraction, is precisely the one that minimizes the time required for the light to pass from one point to another. In fact, by means of an integral, we can associate a time with each possible path, just as we can associate a length with each possible path in space. One may say that light chooses the path that permits it to arrive at its destination most swiftly.

In this form, Pierre de Fermat's principle yields all the laws of geometric optics and, in particular, the laws that French tradition associates with Descartes, even though they are earlier than his work. Because of Descartes and because of his way of regarding light, which dominated scientific thought about light for too long, Fermat's principle was badly understood by contemporaries, whose interpretation of the index of refraction was erroneous. They thought that light was propagated faster in a transparent medium than in a vacuum, and consequently, they inverted the terms of the quotient that defined that index. Only Christian Huyghens appreciated Fermat's principle and used it in his own deep work. It was only at the beginning of the nineteenth century, with the work of Augustin-Jean Fresnel on light waves, that Fermat's work was accepted as a correct fundamental principle.

The calculus of variations is already implicit in Leibniz's thought, which might be paraphrased (following Voltaire) as the pursuit of "the best of all possible worlds." But what is the precise meaning of "the best"? It was not until differential calculus techniques for integrals were formulated in the 1790s by Joseph Lagrange that calculus of variations could be applied rigorously to mechanics. The Newtonian problem of mechanics became a different question of deducing the minimum, or optimum, or extremum of an integral (involving mass velocities, forces, etc.) of a differential system. This was Lagrange's principle, which is now applied in different contexts, mainly under the different names of *optimization*, for example, the Kuhn-Tucker optimization, or the "programs" or "programming" studied by economists.

Lagrange, and subsequently William Hamilton, in a more general form, showed that the motions of bodies described by the equations of Newtonian dynamics are exactly the same as those derived using the calculus of variations to minimize or find the largest or smallest values of a quantity known as the *action*, which depends upon the class of motions being considered. This translation of Newtonian dynamics is known as the *principle of least action.*[1] Toward the middle of the eighteenth century, Pierre Louis de Maupertuis quite confusedly adopted another point of view and showed that, with energy taken to be constant, not the *motion* but the *paths* described by a dynamic system minimize another quantity (not unrelated to the former): the Maupertuisian action. He asserted in 1746 that, "if any change occurs in nature, the quantity of action necessary to produce this change is the least possible."

Since that time, a strange observation has imposed itself: a single theory, be it Newtonian dynamics or classic mechanics, may be written in terms of forces that are either apparently causalist or apparently teleonomic in terms of actions. Indeed, what can be more finalistic than Fermat's principle and the optimization of time as functional criterion? This is no coincidence. In fact, in 1865, Maxwell extended the concept of action to the electromagnetic field and recognized the general character of that concept. His interpretation of light as a form of electromagnetic radiation provided a firmer basis for

Figure 11. *Upper left and right,* Isaac Newton (1642–1727), creator of classical mechanics, and Gottfried Wilhelm von Leibniz (1646–1716), formalizer of the principle of least action. From Bibliothèque Nationale, Paris. *Lower left and right,* Joseph-Louis de Lagrange (1736–1813), creator of analytical mechanics and co-creator of the calculus of variations, and Ludwig Boltzmann (1844–1906), inventor of statistical mechanics. From Bibliothèque Nationale, Paris, and Académie des Sciences, Institut de France, Paris.

Fermat's principle and connected it to the more general principle of least action. By virtue of the concept of entropy, thermodynamics and statistical mechanics in their modern forms have also become susceptible to a strictly analogous formulation.

Recently, mathematicians have begun to study apparently causal differential systems, which may be derived from a variational principle and, therefore, have a "finalistic" character (in the sense as in chapter 1). They have shown that there is a large class of systems with this general abstract property. For no less general reasons, the system that possesses this property may represent real physical phenomena. The general character of variational principles thus emerges clearly; depending upon how one looks at things, every physical theory may be considered both causalist and teleonomic, even though it remains a single theory. At this level of understanding reality, and in this context, causality and teleonomy appear as concepts—actually, rather naive concepts—and we shall return to them.

The calculus of variations can be shown to be closely related to the concept of group theoretical invariance. The action of the variational principle of mechanics is not affected by the application to it of a transformation group, and therefore it follows that the differential equations that derive from the variational principle, namely the Lagrange equations, are also invariant. More precisely, this invariance may be seen by inspection of the integrand, which defines the action and which is called the *Lagrangian* function in honor of a great mathematician and one of the greatest mathematical physicists of all time. Any physical theory that is essentially a field theory, whose prototype is the electromagnetic field, always depends upon an action, that is, upon a Lagrangian function, and the invariance of the Lagrangian function under a group determines the invariance of the field equations that result from it.

There is an even deeper connection between the calculus of variations and group theory; Emmy Noether (1971) deduced from the invariance of a variational principle with respect to a group the identities called *conservation laws*, and this deduction led to a growing awareness that the invariance of a theory with respect to a group determines its suitable physical quantities. Thus, to each degree of freedom of the Galilean group there is a corresponding fundamental quantity of classic mechanics. For example, to a translation in space, there is the corresponding momentum or quantity of motion, $P = mv$. To a translation in time, there is corresponding energy. It is thus not the velocity vector—or its derivative, the acceleration vector—that is important, but the product, P, of the velocity vector by the mass. It might seem trivial to rewrite Newton's classic equation, $mg = F$, as $dP/dt = F$, but from the point of view of statistics, this is, as we shall see, very significant.

It is appropriate to render homage to some of the scientists and philosophers quoted in this chapter: Isaac Newton, Gottfried Leibniz, Louis de Lagrange, Ludwig Boltzmann, Max Planck, Elie Cartan, Albert Einstein, and Paul Dirac. Their portraits are reproduced in figures 11 and 12.

IV.2. Relativity

The theory called *relativity* was the conceptual result of a conflict between two groups: the Galilean group was seen as the invariance group of classic

Figure 12. *Upper left and right,* Max Planck (1858–1947), inventor of the notion of quanta, and Elie Cartan (1869–1951), formalizer of group theory and contributor to the development of the general relativity theory. From Bibliothèque de l'Institut de France, Paris. *Lower left and right,* Albert Einstein (1879–1955), creator of the theory of relativity, and Paul Dirac (1902–1984), main contributor to quantum mechanics. From Bibliothèque Nationale, Paris, and Académie des Sciences, Institut de France, Paris.

mechanics, and the Poincaré group was seen as the invariance group for electromagnetism. Historically speaking, the theory owes its inception to Albert Michelson's negative experimental result.

Let us look at the content of physicists' reflections around 1900 regarding electromagnetism, concepts developed by heuristic reasoning but quite rapidly shown to be entirely erroneous. Various experimental results obtained in the nineteenth century led to a supposition that an ether at absolute rest filled all space, did not necessarily move at the same velocity as matter, and was the medium for the propagation of electromagnetic waves. Among the experiments was that of Hyppolite Fizeau on the velocity of light in a transparent medium in motion.

The model of an immobile ether inevitably implied that the velocity of light measured by an observer who is moving relative to the ether depended upon that relative motion and, in particular, upon its direction. If c is the velocity of light with respect to an immobile ether, and if v is the observer's velocity in absolute value, that observer ought, according to classic kinematics, to observe a velocity of $(c - v)$ or $(c + v)$, depending upon whether he is moving in the same direction and sense as the light or in the contrary direction and sense. An observer who, a priori, knows nothing about his motion with respect to the ether should be able to learn about it by experimentation, by sending rays of light in all directions and measuring the times it takes the signals to reach points on a sphere centered at the source of the signal. If there is motion with respect to the ether, the "ethereal wind" should blow the signal in such a fashion that it reaches first the point of the sphere opposing the sense of motion and, last, the point corresponding to the direction or sense of motion.

This was the principle of a famous experiment by Michelson, through which he sought to demonstrate Earth's motion with respect to the ether. In the Copernican reference frame—which uses the solar system's center of gravity as its origin and the stars' fixed directions as its axes—the velocity of the center of Earth's gravity in motion along its trajectory is about 30 km./sec.; within six months, the direction of its velocity vector is nearly reversed. It may occur at a certain moment that the unknown motion of the Copernican reference frame with respect to the notorious ether cancels the motion of Earth with respect to itself, but such a coincidence could not continue for six months. Using a light inferometer, an apparatus sophisticated for its time, Michelson should have been able to detect an ethereal wind of as much as 2 km./sec. In fact, however, within the limits of precision of his measurements, he was not able to observe any such wind, and he continued to obtain this negative result throughout the year. More recent experiments, from A. S. Kennedy's in 1926 to C. W. Ives's in 1938, all confirm this result. Thus experiments led one to believe that the velocity of light does not depend upon the motion of the observer.

Around 1900, many artificial hypotheses were proposed to explain this negative result, but they often were not consistent with known experimental results. It remained for Lorentz and Einstein (1904–5) to take the only consistent step toward the theory of special relativity. Taking the result of Michelson's experiment as their point of departure, they showed, in effect, that the velocity of light is the same with respect to all Galilean reference frames constituted by classic time and spatial reference frames — defined approximately, over a short interval of time, by the positions of a reference frame related to the center of Earth over the course of its orbit. They thus obtained the principle of the constant velocity of light or of any other electromagnetic radiation: With respect to Galilean reference frames, the velocity of light in a vacuum is always the same and is always equal to c.

There might have been some hesitation at basing a principle of such generality upon the result of a single type of experiment that some other type of experiment might contradict. In fact, Michelson's experiment had only drawn physicists' attention rather powerfully to a well-established, but neglected, mathematical fact, the conflict between the invariance groups that were the basis of classic mechanics and those that were the basis of electromagnetism. The introduction of the ether and the reintroduction of the concept of absolute rest (which, in fact, do not appear in classic mechanics) were the manifestation of that contradiction. Another sign was the impossibility of constructing a consistent theory of the electrodynamics of charged bodies in motion.

To resolve this conflict, Einstein proposed admitting the principle of the constancy of the velocity of electromagnetic radiations, by retaining Maxwell's theory of electromagnetism, which was considered rigorous, and modifying classic mechanics to bring it into agreement with electromagnetism, that is, to substitute for it another mechanics, *relativistic mechanics*, which could be expressed by an invariance group (which was, in fact, the Poincaré group). Classic mechanics, then, appeared only as an approximation — a very good approximation of velocities on the human scale — of the relativistic mechanics being created. Conversely, one can describe relativistic mechanics as the deformation of classic mechanics with a deformation parameter of $1/c$.

Thus Einstein's relativity principle was substituted for Galilean relativity. No physical — in particular, no mechanical, electromagnetic, or electrodynamic — experiment performed in a Galilean frame of reference may permit a demonstration of the motion of one Galilean reference frame with respect to another. The concept of an ether thus lost all meaning, and any theoretical role was abandoned.

Each reference frame carries with it its own time, and the Newtonian concepts of absolute space and time vanish. It is upon the geometrical concept of space-time, introduced by Hermann Minkowski in 1907, that physics would henceforth be based, since the splitting of space and time would from then on be relative to each Galilean reference frame, just as vertical and horizontal axes divide ordinary space into a plane and a line. Henceforth, it was the Poincaré group that would transform Galilean reference frames into each other, and the equations of physics that assume Galilean reference frames would be invariant under the action of the Poincaré group (considered as a group of transformations of Minkowski space-time, which establishes its geometry).

Four-dimensional Minkowski space-time differs from Euclidean space of the same dimension in only one important respect: instead of the square of a vector being the sum of the squares of its components (as is the case in a Galilean frame of reference), the square of a vector appears as an algebraic sum of three squares, with one sign, and a fourth square, which corresponds to time and which has the opposite sign. In this geometrical conception, temporal and spatial directions are distinct and separated by a cone, whose generators express a propagation at velocity c. The Poincaré group, still taken to be ten-dimensional, may be interpreted as the group of motions in space-time that conserve a distance, in a generalized sense. Technically, the Poincaré group is a deformation of the Galilean group, with a parameter, $1/c$; conversely, the Galilean group is a contraction of the Poincaré group, $1/c = 0$.

It is upon the Poincaré group that all contemporary physics is formulated. Everything that we have just said about the application of the calculus of variations to classic mechanics may be applied, without modifications, to relativistic mechanics, including the causalist and teleonomic aspects of the theory.

We must add one qualification, which is hardly technical. No interaction may be propagated with a velocity greater than c, and there cannot be, as there would be for Newton, an instantaneous action produced from afar. Two events, points in space-time, cannot interact if the direction that joins them is spacelike. Any particular theory is required to respect this condition of spatial noninteraction, and it is this condition that one calls, in relativity, the *causality condition.* In a strictly literal sense, two spatially connected events cannot interact with each other. Thus, in a relativistic context, the term *causal* acquires a very precise meaning.

It is appropriate to substitute a relativistic, and thus causal, theory of gravitation for the Newtonian theory. This was the objective of the general theory of relativity, which, as we shall see, was rather badly named. Indeed, it has to do with a relativistic theory of gravity, a theory that Einstein began to develop in 1915. The difficulties of the project should not be underestimated: the Newtonian theory of gravity seemed to be the scientific theory justified by celestial mechanics. While the special theory of relativity was created by Lorentz, Einstein, and Minkowski (between 1905 and 1907), it was to Einstein's great credit that he dared, from motives of consistency, to confront Newton's theory. The observational deviations from Newtonian predictions were minute, and the intellectual effort required to revise his theory was enormous. Aided by the work of two great Italian differential geometers, Gregorio Ricci and Tullio Levi-Civita (1900), and by the applications to classic mechanics that they had developed, Einstein met the challenge.

To see what type of explanation of gravity Einstein arrived at, we must have recourse to a rather accurate analogy that Einstein himself suggested. Let us suppose that Earth is flat, rather like a map, and let us observe the transatlantic courses of ships and planes joining, say, London and New York. If we mark their successive positions on the map, we see that their courses strongly curve north. A further observation, of the shipping lanes in the South Pacific, would suggest the following law: the directions north and south exert an instantaneous attraction from a distance upon ships and planes, and they produce course deviations that can be calculated. This is the type of explanation Newton gave for gravity. But one day, it was recognized that Earth's surface is in fact curved and that, on this curved surface, ships follow the shortest path, or the *geodesic*. This is an Einsteinian type of explanation.

Furthermore, the fundamental phenomenon of gravity is the interdependence of the motions of bodies. To describe that interdependence, Newton availed himself of a scheme of attracting forces and, in fact, obtained an admirably precise description. To describe that interdependence, Einstein avoided such a scheme of forces and wrote field equations that imply that the masses themselves (or more generally the distributions of their energy) curve space-time, each mass imposing its own gravitational field upon a common, exterior gravitational field. It is this imposition, this necessary patching over, that creates the interdependence of the motions of masses. Despite this radically different concept, the theory justifies the Newtonian approach, explains its extraordinary success, and refines its observational predictions only slightly at the scale of celestial mechanics.

We quote below Elie Cartan (1952), who contributed so much to the development of general relativity and who was, in this domain, the most farsighted of scientists:

> One knows that, by the theory of general relativity, Einstein overturned the concept that had been held until then of a homogeneous space-time, preexistent to phenomena and in which phenomena are inserted without altering it. . . . The theory of special relativity respected that concept, and its maximal result was a change in the nature of the fundamental group of the universe. This was no longer the case for general relativity, for which the geometric properties of space-time are in some sense *contingent* and depend upon the distribution of matter. . . . Does

> this imply that the concept of a group has been excluded from physics? Not at all. Quite to the contrary, because the fundamental hypothesis of Einstein's theory is not, as many people thought, that it is possible to formulate the laws of physics in any arbitrary system of coordinates, which would be an elementary *tautology*, but that in any sufficiently small region of space-time the laws of special relativity are true up to a first approximation. (9–10)

For this reason, the Poincaré group continues to play its fundamental role. The theory of general relativity thus does justice to the theory of special relativity, which remains valid in the neighborhood of each point in space-time. It implies a causal theory for the gravitational field, which, like the electromagnetic field, is propagated by waves traveling at a velocity of c. Gravitational wave fronts are geometrically identical to electromagnetic wave fronts, since gravitational rays coincide with electromagnetic rays: both take the form of geodesics in a space-time tangent at each point to the cone that defines the separation between the spatial and the temporal. Gravitational waves, which are an integral part of the theory, have never been observed because of the very weak relative energy they are apt to carry. Einstein's equations derive from a variational principle corresponding to a Lagrangian function, which, in a vacuum, is nothing other than the (scalar) curvature of space-time itself.

Thus one sees that, even though relativity introduced a necessary revolution in our concepts of space and time, it was nevertheless cast in the general shape of the physical theories described.

IV.3. Statistical Mechanics

From Ludwig Boltzmann and Josiah Gibbs we have learned that a fluid is composed of a great many particles or molecules in motion. This is according to elementary laws of mechanics, which may be adopted, for the moment, for classic mechanics. And we have seen that, for reasons connected with the Galilean group, the "suitable quantities" of classic mechanics are momentum and energy, whose sums are conserved in the collisions between two or more particles.

We consider here only the simplest case, a fluid whose particles are identical to each other. Given such a fluid—the characteristics of whose component particles are known—a state of that fluid at a given instant is yielded by a function that describes the number of particles in a given, infinitesimal volume and whose momenta differ infinitesimally. Thus we have a distribution function, f, which depends at every instant upon the position and momentum under consideration.

The principal problem in statistical mechanics consists of reconstructing the macroscopic state (density, mean momenta, energy, pressure, viscosity, etc.) of a given fluid from its microscopic description by a distribution function and of deducing the changes in time that the macroscopic state undergoes from the evolution of the distribution function. Coarsely stated, it is a question of extracting from a delicate, microscopic description information on a macroscopic scale, information impossible to obtain by theoretical methods while remaining within that macroscopic scale.

Turbulence and statistical mechanics in the broad sense follow the same general scheme and differ only in their orders of magnitude and technical points; those differences are frequently due to the different historical development of the fields. All general analyses regarding turbulence remain valid in the domain of classic statistical mechanics, even if the orders of magnitude of the phenomena under study are not entirely the same.

It is, of course, useful to analyze one of the technical considerations that lead to different approaches. To pass from microscopic to macroscopic scales, one proceeds by averages calculated on the basis of a distribution function. But averages of which quantities? Depending on the choice of microscopic quantities, one obtains different fundamental macroscopic quantities. We must remember a significant mathematical fact: while the mean of a sum is the sum of the means of the terms of that sum, the mean of a product of factors is not the product of the means of those factors (see I.4).

The density of a fluid appears as the mean of the number of particles within an infinitesimal volume, if the atomic weight of a particle is taken as the unit. This is the starting point for both turbulence theory and statistical mechanics. The next step, however, may differ. Statistical mechanics rigorously defines the *mean momentum* of the fluid as the average of the sum of the momenta of the elements; it proceeds in similar fashion for the *energy* of the fluid. Working with the mean momentum and a density, one is led to define the macroscopic velocity vector of the fluid as the quotient of its mean momentum vector divided by its density. Turbulence theory generally defines the macroscopic velocity vector directly as a mean velocity vector, and because of the aforementioned mathematical fact, the theoretical results will then be different. Similarly, pressures and viscosities will be deduced differently from the same microscopic data.

One school of thought, inspired by Alexandre Favre (1958; 1965a; 1969; 1992; and Favre et al., 1976), on the contrary, uses, and quite properly, the same macroscopic examination grid for turbulence used in statistical mechanics (see II.4.2.2). The choice adopted for such a classic statistical mechanics offers the advantage of passing easily into relativistic mechanics, since the Poincaré group makes a momentum vector of space-time (whose temporal component relative to a reference frame yields the corresponding energy) obviously appear as a suitable physical quantity. Momentum and energy are synthesized in a single geometric object.

According to the macroscopic grid by which one examines a single microscopic datum, the theoretical results — the equations of motion, for example, or the possible fluctuations of that datum — may be different. But both of these apparently different theories interpret all the experimental data correctly and reasonably, up to the same approximation. Further, the preceding facts and the importance of the macroscopic examination grid are not yet fully recognized by some physicists.

Statistical mechanics permits us to expose the fluctuations of macroscopic quantities (also called *observables*) of microscopic origin and to estimate their upper bounds — but not to predict them. The delicate analysis of entropy by the Boltzmann theorem (also called the H-theorem) demonstrates the role of these fluctuations. They are functions of the molecular agitation described by temperature. At ordinary temperatures, there appears a background noise, which is a sign of that agitation; this noise systematically distorts observations, even at macroscopic scales, and becomes relatively negligible only at very low temperatures, close to absolute zero. This is why certain experiments that require great precision are performed at very low temperatures.

A microscopic datum, while itself perfectly deterministic in the usual sense of the term, gives rise, by a change of scale and by a procedure of means and deviations from those means, to a theory said to be *statistical* and apparently no longer deterministic in the previous sense. Physics has increasingly concentrated on a theory of measure for the values of observables in certain states: quantities with numerical values, accessible to experiment, and that one takes seriously. The postulates of Laplacian determinism (if one knew perfectly the positions and velocities of bodies at a given instant . . .) appear to derive, even in the classic domain, from a philosophy out of touch, except ideally or even ideologically, with the realities of physical systems (see I.5.2).

IV.4. Quantum Mechanics

The prehistory of quantum mechanics is dominated by Planck's hypothesis regarding the theory of radiation, which, by assuming discrete levels of energy, sought to eliminate an apparent contradiction between Einstein's remarkable result concerning the photoelectric effect (1905) and the work of Niels Bohr seeking to describe phenomena observed at the atomic scale by means of an atomic model resembling a solar system — clearly an indefensible model even though still of great heuristic value for experimental physicists. This was called the *first quantum mechanics.*

The experimental basis for quantum mechanics, which was created by Louis de Broglie, Werner Heisenberg, and Erwin Schrödinger, was provided by spectroscopy. Atoms emit or absorb energy in the form of electromagnetic radiation at special frequencies that identify them but that present certain patterns. How could these patterns be explained theoretically? And, on the other hand, radioactivity had just been discovered, quite stranded, without theory.

For a long time, mathematical physics had at its disposition a theory of vibrations or frequencies (eigenvalues) that, in the domains of mechanics and acoustics, gave rise to sets of frequencies, or spectra. The most elementary example was offered by the theory of vibrating strings. More generally, a theory of eigenvalues for matrices or operators was proposed and subsequently exploited by Heisenberg and Schrödinger to explain the spectra of atoms. It remained for Paul Dirac and his brilliant work, which was completed by Wolfgang Pauli, to stimulate the powerful synthesis constituted by relativistic and nonrelativistic quantum mechanics.

Relativistic quantum mechanics, in a broad sense (i.e., including quantum field theory), currently offers the double and curious quality of both being based on mathematical concepts and being mostly inconsistent from a mathematical point of view. While the different fields, or particles (actually, what we call a particle should be considered a field), are connected with representations of the Poincaré group, there are many things we do not know, and we lack a consistent mathematical structure within which to place these phenomena. But this theory offers us such an apt instrument for the understanding and control of the microscopic world that it seems to be, simultaneously, the most fundamental and most unsatisfying of physical theories.

Different, but more or less equivalent, ways of presenting quantum mechanics in the broad sense have been devised, but few of them touch the imagination. But we must be wary of those few that do, because they carry with them images that are human—all too human. It may be in terms of deformations that quantum mechanics can be best presented—and with a minimum of mutilation.

Let us begin from the classic mechanics of a system of n particles. For an arbitrary n, we are led to offer a description of the dynamic geometry of the atomic system in terms of phase-space, a description that goes back to Hamilton and even to Lagrange (around 1790). A dynamic state of the system is described by the set of the positions and the momenta of all the particles, that is, by a point in a more or less abstract, even-dimensional, "space," which is precisely the phase-space. This space is automatically equipped with an additional geometric structure, called the *symplectic structure* because it is related to the symplectic group, which endows it with a volume element and permits a definition of a composition law between two functions of the phase-space. This law, which requires a change of sign when those functions are interchanged, is called the *Poisson bracket.*

For a physical system described geometrically by its phase-space, the dynamics, strictly speaking, is determined by the energy function H, or Hamiltonian. The fundamental equations of classic dynamics are restated faithfully by saying that the change in time of any function, u, or any function that is observable upon the phase-space, is yielded by the Poisson bracket of that function with energy, H. One thus obtains an intrinsic description (without coordinates) of classic dynamics, where one still has the usual product of two functions of the phase-space, uv.

What happens when we deform the two composition laws for functions of the phase-space in a consistent fashion as a function of one parameter? As usual, the initial product deforms into a product always associative but not commutative, and the Poisson bracket deforms into another bracket with similar properties. Using this method for a simple type of deformation, with the parameter taking the value of Planck's constant, h, one obtains a form of nonrelativistic quantum mechanics. This constant, which has the dimensions of an action, figured in the earliest work on radiation in order to render its energy discrete, or quantified. This point of view (which was introduced in 1932 by Eugene Wigner and Hermann Weyl in another context suggested by classic statistical mechanics) is susceptible to broad generalization, especially for relativistic quantum mechanics, and has in recent years stimulated many important theoreticians.

The observables of the system are described here by functions of the phase-space (positions, momenta, energy, etc.), and they obey a dynamic equation that is the deformation of the equation of classic dynamics. The states are described by a statistical, generalized function, or Wigner's function, P. The expectation value of an observable, u, for a state, P, at a given instant (the only object that can be measured) is nothing other than the integral, $\int Pu$, over the phase-space. Like a probability density, Wigner's function is normed, with the integral equal to 1, although it need not be positive. It satisfies a dynamic equation analogous to the one satisfied by the observable. Such a function exists in classic mechanics, but it differs from 0 at only one point in the phase-space, the point that indicates the position and momentum of the system under consideration. In the quantum case, the function is bounded by the inequality, $|P| \leq (2/h)^3$. Since P has an average equal to 1, it must be different from 0 in a domain within the phase-space of measure at least equal to $(h/2)^3$.

This formulation is within the context of the formulation of Heisenberg's uncertainty principle. Position and momentum are inextricably connected, just as in relativity time and space are necessarily connected in any change of reference. There is no state for which simultaneously, perfect measurements of the position and momentum of a particle can be performed. But only measurements are accessible to us. The theoretical change of observables over time is entirely deterministic, as is that of Wigner's state function; but as in statistical mechanics, it is impossible to deduce from that determinism a simultaneous, precise measurement of position and momentum.

The lower bound of the volume delimited by $(h/2)^3$ for P is equivalent to $\Delta p \Delta q \geq h/2$ for a particle (the classic form of Heisenberg's uncertainty inequality) obtained in terms of a wave function—or a probability wave—introduced to describe the state of a system in another context. Thus, if the lower bound is reached at a given instant (a minimal wave packet), it cannot be attained again at a later instant. Further, the fundamental equations of quantum mechanics, like those of classic mechanics, derive from a variational principle: the optimization of a certain action. Quantum mechanics is thus apparently a statistical theory, or if one prefers, it has a statistical form but is not statistical for any definable set of events.

One might think, following great scientists like Albert Einstein and Louis de Broglie, that quantum theory is an incomplete theory and that there is a possibility of completing it so as to make it a true statistical theory. This idea has led to numerous attempts to develop what was called a *hidden-variable theory*, one whose variables would describe a submicroscopic level whose wave functions, or Wigner functions, would yield the statistics.

In 1969, John Bell established the inequalities that concern the experimentally accessible quantities that distinguish quantum mechanics from the hidden-variable theories that complete it and that satisfy certain reasonable hypotheses (locality, separability, etc., which are too technical to be stated here). Delicate experiments have been performed to test Bell's inequalities, the last and indeed only really decisive ones having been performed by Alaine Aspect and G. Roger Dalibard (1982). The experiments favor quantum mechanics in its traditional form and militate against hidden-variable theories. If one insists on supposing the existence of hidden variables, one must deny the general and reasonable hypotheses that Bell introduced. We would thus lose much more intellectual comfort than we would gain, and we would find ourselves confronted by a much stranger physical world. This seems to have discouraged further attempts at constructing such a theory.

Such is the situation. We must emphasize that what the contemporary physicist calls a *particle* in quantum mechanics has no relation to our intuition of a small, localized fragment of matter and energy. The fundamental physical objects about which we reason are really *fields* more or less modulated in space-time. No paradoxes stem from quantum theories or from the experiments that routinely support them. Only certain vulgar modes of thinking about the localization of a photon or the possibility of splitting an electromagnetic field into two or more photons prove themselves completely inadequate, and it is they that suggest paradoxes.

Thus, for physical theories, for classic mechanics, and for quantum mechanics, the concepts of cause and effect, as the ancient formulated them, have become too naive, too related either to our perceptions of real phenomena or to the way we undertake rational processes. Science has in fact developed into a tightly knit network of interactions, and access to that network may be achieved by steps whose order carries with it a certain degree of arbitrariness, although we may to a certain extent choose the set of assertions we will use as axioms, whether in mathematics or in theoretical physics.

It would seem that our sense of causality or teleonomy derives in part from the choices we make — choices that express our taste, our esthetics, our economy of thought — and these evolve with time. It is true that these partially subjective views of science carry multipurpose heuristic burdens that, at every moment, can stimulate our imagination.

Note

1. It is customary to suppose that the mechanical systems under consideration are subject to holonomic constraints and are frictionless. In fact, macroscopic dynamic systems that describe real phenomena may comport nonintegrable (nonholonomic) constraints on their momenta or friction. In the first case, one can find a somewhat more elaborate form of the principle known as least action, where there appears an extremum of the action subject to constraints, a principle equivalent to the differential equations of the dynamic system. Such is not true of the second case, but the macroscopic description of friction is only a phenomenologically coarse synthesis of surface phenomena that occur on a different scale, a scale in which they may be described by holonomic differential systems.

V
Biology

It goes without saying that in biology we arrive neither at the simplicity of fluid flows, which have an internal order disguised by turbulence, nor at the lucidity of statistical and quantum mechanics. The methods of reasoning and techniques of discovery in biology may seem rudimentary or even simplistic in comparison to mathematics, which is pure, nearly disembodied, science. Mathematicians even refuse, on occasion, to admit that biological investigations are indeed science, and certain mathematicians do not accept the validity of the inductive method—which however has been used successfully in biology. But the methods of biological science are in fact not so different from those of mathematics, and the deductive method that mathematicians consider the unique source of discovery is not far from the arguments of biology, even if biologists are frequently obliged to resort to less than perfect arguments.

In spite of the occasionally brilliant gropings of eighteenth-century biologists, experimental biology was not truly launched until after 1800. Its development in the twentieth century was impressive and led to prodigious advances in medicine in the last fifty years of the century. Biological researchers now use the most recent techniques and instruments developed by other disciplines, from sophisticated electronic microscopes to nuclear magnetic resonance scanners to probability theory and computer science. It is in biology that many of these modern techniques find their final applications.

In the spirit of this collective study of philosophy and science, the two ideas we discuss here are *teleonomy*, which leads us to causal embryology, and the *double helix* of ribonucleic acids, which leads us to molecular biology, with its pronounced orientation toward physics. It is in these two branches more than in any other domain of biology that we are confronted by these two ideas.

V.1. Teleonomy in Embryology

One is constantly tempted to say that everything happens as though such and such a process had such and such an end in view, or as though there were a convergence of reactions toward a goal, or as though the entire being determined the nature of its parts. So why not say so explicitly? Is teleonomy a mirage to which we are always susceptible, or is it a reality we cannot escape?

Several recent articles deal with teleonomy, and the idea has been resurrected by several points of view. Philosophers in particular are interested in it, as well as some scientists—especially physiologists. A recent book deals with the question in a rather fragmentary form and without much real coordination (Parrot, 1985). It might be useful to look at the question from the point of view of an embryologist, because it is in that discipline that one encounters the most characteristic phenomena and those that are arguably the most teleonomic (Wolff, 1983, 1984).

What is teleonomy? It is in Jules Lachelier's book (1924) that we find the first consistent definiton of the *efficient causes* and *final causes* of teleonomy:

> The conception of laws of nature seems to be founded upon two distinct principles. According to the first, phenomena form in a series in which the existence of each element determines that of its successor. According to the second, these series form, in turn, systems in which the idea of the whole determines the existence of the parts. Now a phenomenon that both precedes and determines another is what has always been called an *efficient cause*, and one that *produces the existence of its own parts* is, according to Kant, the true definition of a *final cause*. (11–12)

We detect a certain hesitation in Lachelier's argument and, consequently, in the definition he proposes. It is not the "idea of the whole" that determines the existence of its parts. It is the whole itself! There is reason to refute an excessively broad interpretation of this definition. One might think that, in this sense of teleonomy, it is the future that in some sense determines the present. The *whole* is an, as yet, inchoate and enigmatic set, a *possible reality*, if we may express ourselves in such apparently contradictory terms. We would like to show that, at the very beginning of embryonic development and also further along in the process, certain phenomena, certain properties, appear that correspond well to Lachelier's definition.

V.1.1. There is a stage in the development of most animals and plants where nothing is yet determined. *Determination*, for the embryologist, refers to the precise stage of development preceding *differentiation.* Differentiations are not yet visible, but they are already in place. Before the stage of determination, the location and nature of organs are not yet fixed. Everything is, in some sense, indeterminate: one might even say uncertain, vague, loose. Below we give examples of such indeterminacy among both vertebrates and invertebrates.

Two kinds of egg have long been distinguished, mosaic eggs and regulation eggs, and this distinction is noted at the earliest stages of development. But this distinction has been weakened by observations that mosaic eggs are capable of a certain degree of regulation and that regulation eggs display, sooner or later, a mosaic of determined parts. The classic example of the regulation egg is that of the urchin. Let us consider the 2-, 4-, 8-, and 16-blastomer stages: at each of these stages, any cell has the possibility of developing into a complete embryo and larva. Prototypes of the mosaic egg are the eggs of the ascidian and the mollusc. The destruction of one part of the egg or of several cells of the embryo has the effect of depriving it of one or several organs. But in this case, as well, it has been shown that some regulation is possible.

Recent experiments performed on vertebrates show that the phenomenon of regulation has great scope. Mammalian embryos may have the greatest capacity for regulation; a human ovule can sometimes divide into as many as three or four distinct embryos. The phenomenon is constant in the armadillo, one species of which always bears four individuals and another species of which always bears eight or twelve. This twinning occurs late in gestation, that is, not at the stage of the first divisions of the ovule but at the stage of blastoderms, which are already composed of thousands of cells.

Hubert Lutz's experiments on birds confirm the multifarious potential of the blastoderm: he separated birds' germs into two, four, or eight pieces without any particular spatial orientation and obtained, in principle, as many embryos as he had separated fragments. These embryos were complete as far as could be determined, since they generally did not progress beyond the stage when the primordial organs (the nerve trunk, the dorsal cord, the digestive tract, the sense organs, etc.) differentiate.

Eggs of amphibians (frogs, tritons, salamanders) have properties that lie between those of birds and molluscs. The fertilized frog egg is notable for certain features very early in its development. Soon after fertilization, there appears a gray crescent, which fixes, even before the segmentation of the egg, into distinct cells: the dorsal side and the bilateral symmetry plane of the future individual (Ancel and Wintemberger, 1948). If one separates the first two blastomeres with a slipknot, there are two possible outcomes. (1) The plane of separation of the two blastomeres coincides with the plane of bilateral symmetry, and each of the blastomeres gives rise to a complete embryo. (2) The plane of separation of the two blastomeres separates a dorsal one from a ventral one, and only the dorsal blastomere gives rise to a complete individual. The ventral blastomere develops into a piece of abdomen, "ein Bauchstück," in Speman's terminology (1901, 1902, 1903) (i.e., a set of nondifferentiated cells). There are several possible intermediate stages between an unorganized clump of cells and a clump that begins to show organization; it depends upon how much of the gray crescent the ventral lump contains. This example shows that, in certain eggs, there is a material more or less determined with respect to the dorsal-ventral direction, while there is no determination with respect to the lateral direction.

It may be astonishing that there is such great disparity in the destiny of parts of eggs, that in certain groups of animals nothing is determined at the earliest stages of development while, for other groups, everything (with some qualifications) is already determined.

In the first phase of the development of regulation eggs, the position of the organs is still random and the future state of the anlage is not yet fixed. One can consider an embryo as a circle or a sphere that corresponds to a specific species—human, calf, duck—and that has an axis defining the direction of its plane of symmetry and a polarity corresponding to the orientation of the head. But it is the *direction* of the plane of symmetry, and not the plane of symmetry itself, and the *orientation* of the head, not the position of the head itself. One may recall Pascal's maxim, applied to another aspect of reality, "It is a circle whose center is everywhere and whose circumference is nowhere." What is remarkable is that, in the beginning, nothing is defined or fixed except these abstractions. There is a preestablished plan or program, nothing else. We even hesitate to use the term *preestablished* for a clump of matter that is still so malleable. In another vocabulary, one could speak about the *idea* of a certain organism or species, but such a formulation would tend to support a restrictive concept of reality.

The results of many experiments performed upon the embryos of birds, mammals, and many other species show decisively that a head may develop at the site of the trunk or the tail, a complete individual may develop from a lateral half or a posterior half, and one organ may develop at the site of another or in addition to another. A single entity may give rise to two, four, or eight complete individuals. Certain species have not developed that property we call *regulation* as much as other species have. Regulation is more or less important and lasts long or only briefly, but most embryos possess it to some degree—and lose it sooner or later.

Most eggs during the course of their development—some immediately after fertilization and others even before that—exhibit *radial symmetry* (for example, the eggs of amphibians, urchins, and mammals). This means that they are perfectly homogeneous in all spatial directions. Minutes or hours after fertilization they already exhibit a bilateral symmetry. How does the passage from the one symmetry to the other occur? Indeed, how does the passage from the homogeneous to the heterogeneous occur?

The most remarkable model has been proposed by Paul Ancel and Pierre Wintemberger. The frog's egg is heterolecithal, which means that it carries a large quantity of vitellus, or nutritive reserves, that is distributed according to a density gradient scaled between the two poles of the egg, the vegetative pole being loaded with vitellus and the animal pole being richer in cytoplasm. At the moment of fertilization, the egg still adheres to the inert membranes that form its shell. Fertilization has the effect of uncovering the vitelline membrane, thus liberating the egg from its adherence. Consequently, the egg, carried by its vitelline weight, tips, to arrive at an equilibrium position. In this movement, the vegetative pole describes the arc of a circle in a vertical plane, which plays an important role in the egg's future development. This plane becomes the plane of bilateral symmetry of the future organism. The rotation also defines the future dorsal and ventral sides of the embryo. The dorsal side is realized by the unpigmented crescent, the gray crescent, or the light-colored crescent, depending upon the species; the crescent renders the rotation of the egg visible.

One may say that, at this stage and because of this process, the organism passes from being an egg to being an embryo. From this stage onward the most important systems — in particular the nervous system, which is situated on the dorsal side, and the gray crescent, the first appearance of which is also on that side — fall into place. Thus at an early stage, and even before any differentiation occurs, the egg passes from a homogeneous state to a heterogeneous state, from which passage the entire future of the organism is derived.

That gravity determines the plane of bilateral symmetry and that one can situate this plane in any direction one wants, according to the arbitrary position one assigns the vegetative pole, shows that the plane of bilateral symmetry and the complete organization of the egg are determined before the inception of differentiation. The rotation, called *symmetrization*, triggers the materialization of an organization, which is virtual before being localized. Therefore, a detail as trivial as the rotation of the egg according to the weight of its nutritive reserves has profound consequences on the localization of future organs. We do not know what reactions occur among the different particles that constitute the egg in its different levels, but we can say that they are uniform at all the egg's meridians. Yet once a simple, localized, physical stimulus occurs, it can provoke a preference for one of those meridians.

V.1.2. Another property found to a greater or lesser degree in certain species is *regeneration.* It has no direct relation to regulation, even though the result is similar. This is the property that certain organs and organisms possess of reconstituting or replacing the parts of themselves removed by accident or autonomy. Species that manifest the greatest capacity for regeneration are not necessarily those with the property of regulation. Thus the embryos of birds display the highest degree of regulation, although as adults they have very little capacity for regeneration.

Many species of planarians (flat worms) have a great capacity for regeneration, although their embryos have very little capacity for regulation. The essential difference between the phenomena is that in regulation the organism constructs a piece of itself that never existed while in regeneration it reconstitutes an organ or a region that once existed. Regeneration is another manifestation of the teleonomy inherent in an organism. There is nothing more curious than to watch a mutilated individual—a worm of the Tricladida order, for example—reconstitute what it lacks (epigenesis) or recompose itself completely to reorganize a complete individual (morphallaxis). Here again, we can speak of programs, of a directing plan. What is remarkable is that a part of the body that regenerates stops when it has accomplished its mission, even though it could go much farther.

Regeneration is common to many organisms and is produced by processes that sometimes differ from species to species but that operate with the same objective. It generally occurs by two processes.

In the first process, specialized cells, called *regeneration cells*, are exploited. These cells are found in many categories of worm, among them the planarians and the Annelida. In other cases, there are no regeneration cells, strictly speaking, but the different tissues possess cells that, by successive redifferentiations, can reconstitute differentiated organs or tissues. In certain species, all the cells of the organism can contribute to total reconstruction. This is the case for the less-organized species, such as the coelenterata, the echinodermata, and the sponges. What is curious is that certain species possess a great capacity for regeneration and others none. Thus the dendrocoela have almost no power of regeneration, and the planarians have this power to a remarkable degree, despite the fact that the two orders are very closely related.

The second process consists of *induction.* It is the capacity possessed by several organs, differentiated or not, to orient the regenerating cells toward a certain differentiation. Thus in most animals capable of regeneration, the base — that which remains of the mutilated organ — has a determining influence upon the regeneration bud. This is why, in general, the organism does not regenerate more than what it lacks, as in the case of the induction of eyes by the cerebellum of a planarian or of a limb by the nerve column of a triton.

Induction, of which many examples are known, is the mechanism underlying the teleonomic explanation of regeneration. There are always efficient causes to explain an instance of determinism: in other words, there is always a mechanistic explanation for phenomena. But is that enough? These are the material instruments that nature's programs and plans employ. The mechanisms of regeneration are, first of all, successive inductions with inhibition processes. Can we deny that ontogeny is the realization of a project according to a preconceived plan? But the execution of this program is accomplished with considerable flexibility. Is this the contingency within the finalist hypothesis?

V.1.3. When one attaches together two embryos of amphibians or mammals at an early stage in their development, one possible result is a single individual. There are occasionally conflicts regarding the individual's sexual identity, but there is a more or less clear tendency toward unity. This result is still more remarkable because it is not natural — it does not occur spontaneously: we must force the situation, not by violence but by gentle intervention. Around most eggs there is a protective envelope or hard gangue, which must be removed cautiously. One then brings the two exposed eggs into contact. It is by this process that mammalian (mouse) eggs have been fused. Sometimes nothing of their original duality remains. In the best cases, this duality never existed; it was only in our mind.

Let us give another proof of the existence of a wholistic factor. Among the tetrapodous vertebrates, limbs generally have three articulated segments. The rear limbs, for example, have thigh, leg, and foot segments, corresponding to femur, tibia, and tarsus-metatarsus. In experiments performed on chicken embryos, when undifferentiated leg material was added to the normal segments giving the sequence femur + tibia + femur + tibia + foot, the result turned out to be femur + tibia + foot, that is, an apparently normal limb. Conversely, if tibial material was removed so as to produce the sequence femur + foot, the result was still a limb with three bones, femur + tibia + foot. The tibia was resurrected, so to speak, even though it had not existed. But the whole segment still showed a certain anomaly; it did not form a fibula. The fibula, a partner of the tibia and normally a rudimentary part of the skeleton, manifested that rudimentariness by disappearing in favor of the tibia, the principal bone, which appropriated material from the weaker bone. The fibula is usually a rudimentary bone with a single epiphysis.[1] In this case, it evolved into a complete bone with two epiphyses and a long diaphysis.

Embryologists interpret this result by invoking a regulation phenomenon, but more is involved. There is an authority of the whole over its parts, which are really subordinate to the "decision" of the whole. This is, of course, a convenient mode of speech, which, however, ought not deceive us. How are that authority and that subordination established? Obviously, there are causal phenomena at the origin of such differentiations, but those efficient causes are subordinated to a rule imposed by the whole.

Let us return to the example of the limbs of bird embryos. If you add one or two supplementary bones to the normal alignment of articulated segments—for example, a femur plus a tibia, and including, of course, the material that would normally yield all their surrounding soft tissues—they become integrated into the limb. But instead of five articulated segments, there are only three. This is *regulation of excesses.* Conversely, let us remove a limb segment, the one that corresponds to the tibia. There is a regulation; the tibia reconstitutes itself, nonetheless. This is *regulation of deficiencies.*

When there is too much material, the excess is assimilated. When a segment is missing, it is reconstituted. There seem to be certain sequences that cannot be realized. A thigh and a foot, for example, cannot develop in mutual contact: an intermediate segment is reestablished. This is seen in the regeneration of birds' articulated segments and among batracia and insects as well.

It is remarkable that two nonconsecutive bones cannot seem to form or coexist adjacent to each other. They cannot form unless an intermediate segment is interposed between them. How can this incompatibility be explained? It is one of the most mysterious and paradoxical phenomena that can be imagined. Obviously, we do not expect to explain it by invoking mysteries and paradoxes, but we need not refrain from recognizing how extraordinary it is. When the explanation is discovered, it may be based upon an immunological hypothesis, like the explanations for many incompatibility phenomena. Of course, recognizing is not explaining, but recognition always precedes explanation.

Organisms and organs tend to reproduce complete series, to normalize deficient series or series with superfluities, to fill in voids, and to suppress excesses. This can occur only at certain moments, at certain stages, while tissues are still malleable, subject to influences, capable of undergoing induction. This is a remarkable property. Imagine the surprise of the experimenters who, thinking they had, for example, removed all the tibial material from an embryo, during the autopsy found a tibia. One can obtain similar results in several species. These cases are significant, and their implications transcend pure causality.

Another important theme is that there is a preestablished plan, a plan present before its realization. Many experimenters demonstrate the virtual presence and the realization of a plan in various embryos as well as in adult organisms undergoing regeneration. This plan is the same for all vertebrates: general features appear first, details emerge gradually. Development follows from the general to the particular, from the abstract to the concrete.

Etienne Geoffroy Saint-Hilaire (1822) even wrote about the unity of the organizational plan of all animals—an exaggeration and quite inexact. There is an identity of function among most animals, but the organs that perform the function have sometimes only the vaguest similarities—a digestive tube, organs for assimilation and excretion, genital organs, and so on. But within specific groups, such as phyla (large classes of animals), there are not only analogies but anatomical homologies. A plan directs the principal phases of their development; one can even use the term *master plan.*

With these conditions we may associate another, drawn not from ontogeny but from phylogeny: since the secondary era, reptiles have had a tendency to realize mammalian properties. During their evolution, the structures that prefigure mammalian structures occurred three or four times, so that one does not know which are the true ancestors of mammals.[2] Phylogeny offers much less certitude than ontogeny with regard to teleonomy, but living species do exhibit series oriented in certain directions, and that orientation is irreversible. We can give numerous examples; among the reptiles, in particular, there have been several attempts to create a flying animal (dinosaurians, pterosaurians, the archaeopteryx). These attempts to create something new either fell short of their mark or overshot it: consider, for example, the mammoth's curved tusks and the deer's antlers. This is called *hypertely.*

One can cite the existence of monsters as an objection, since they imply a defect in teleonomy. But they do not really constitute a serious objection; regulation is an effect of the common processes of teleonomy, but it can act blindly, without any relation to the end to which it is tending. These are unconscious mechanisms, capable of acting if something is upset in their organization or inception. There are phases, stages in development, where regulation acts to the profit of the organism. Once these stages have been passed, regulation can stutter. Thus monsters like cyclops, symmeli, and anencephali are not arguments against teleonomy.[3] On the contrary, they show that there are certain rigorous mechanisms that lead to consistent results under certain conditions; these constitute the norm.

Figures 13 and 14 show two kinds of cyclocephalus, while figure 15 shows two kinds of symmelus. Figure 16 shows an omphalocephalus after ten days of incubation.

We recall watching close up, through a small glass window, the development of a monster created experimentally as a result of a very early operation on a chick embryo (figure 16). It was an extraordinary monster, with its head (*b.i.* to *b.s.*) inserted into its stomach. Its heart (*C*) was on its neck. It had a double liver (*F*) and an aberrant brain. Such a monster had never been seen later than the third day of incubation. Experimental techniques already permitted creating one at will. Thus we could follow such a monster up to the tenth day, when all the organs of a normal embryo are differentiated and recognizable. When this embryo was removed, we had had a surprise in store: it seemed to be nearly normal despite all its malfunctions. The teacher, Professor Ancel, said, "This is not astonishing. It is seeking to return to normal." This suggestion, made by a man who admitted only a rigorous mechanism, showed that an avowed enemy of teleonomy involuntarily still used a vocabulary that admitted teleonomy.

One is tempted to use the word *preestablished* when speaking of a plan or a program, but if everything is preestablished in a program, where is there a place for contingency? Does not development take place according to a rigid determinism, according to laws of efficient causality and nothing else? Certainly, this explanation is valid. But isn't there more that is teleonomic? The question is extremely embarrassing, and the solution escapes us the moment we believe we glimpse it. If everything is determined in advance, how can we explain that a virtual something can act upon a still nonexistent organism? Have we the right to conclude that a principle, an idea, acts upon the development of future events? We return to the point of departure, that is, the problem of teleonomy. Nothing is materially organized, but everything is prepared. This is the same problem as the opening of a flower bud or the development of a plant seed. Nothing is physical, nothing is present, but the entire future is in germination. The egg is, in some sense, a "pill" of virtual forms and, in the language of embryologists, of potentialities.

The reader may be surprised that neither DNA nor RNA has figured in this discussion and that we have not discussed any of the enzymatic procedures that govern all differentiations. It is because these concepts are universally conceded: they correspond to a deterministic explanation, which is now beyond debate, of biological reality. But one may look beyond and behind that explanation, which is what we have tried to do here, and such a perspective leads to questions that remain unsolved.

It is not without hesitation that we invoke ideas that revive the problem of teleonomy. This concept is out of favor with most biologists, who believe

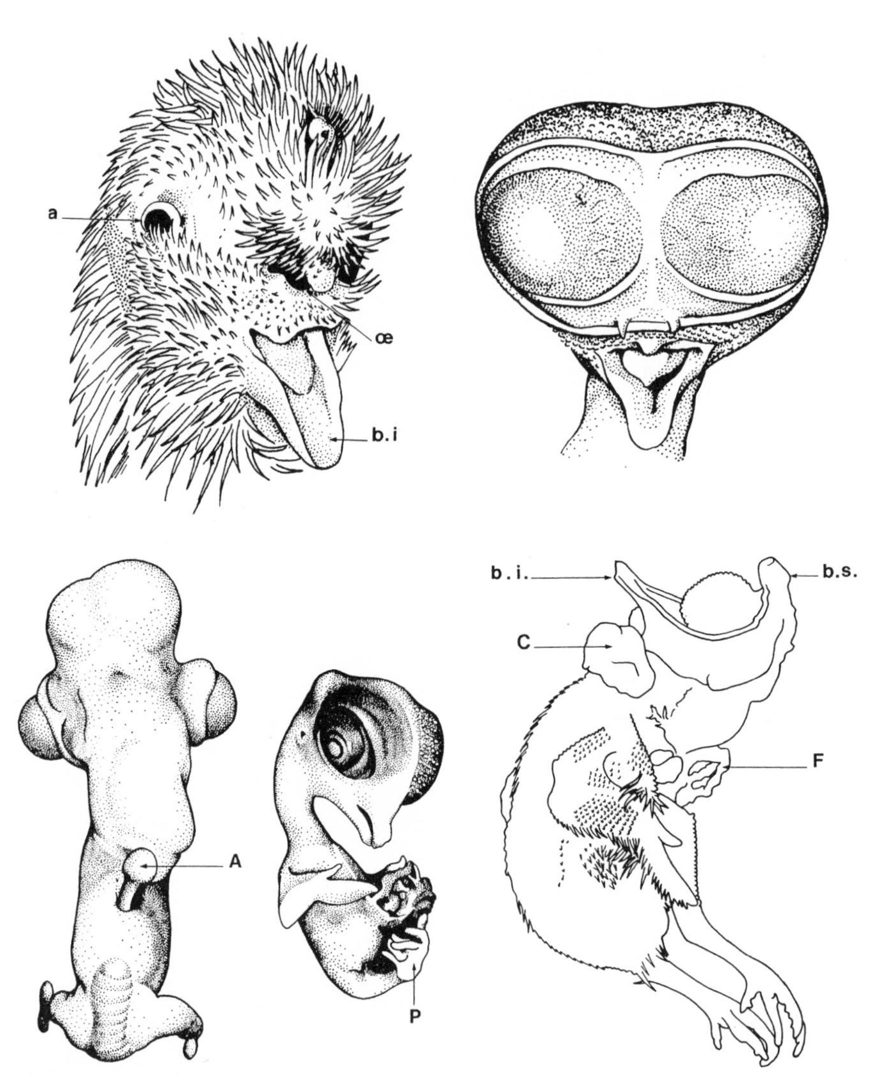

Figures 13 and 14, *upper left and right,* a true cyclops with one reduced eye, *œ*, and external ear, *a*, and an inferior beak, *bi*; and a cyclocephalus with two eyes fused in the middle. From Wolff, 1936. Figures 15 and 16, *lower left to right,* a symmelus with a single wing in the back, *A*; a symmelus with a single leg, *P*; and an omphalocephalus with inferior beak, *b.i.*, superior beak, *b.s.*, and remnant of head between heart, *C*, and liver, *F*—the posterior part has become quasi-normal. From Wolff, 1936 (first two) and Wolff, 1933 (last).

only in rigorous, mechanical laws. It is not a question of denying these laws but of observing that they do not explain everything. Teleonomy does not assert that the future influences the present; the future does not exist. But it may be *prepared* thanks to real systems that presuppose that eventual realization. An entity can preexist its elements. It is in this sense that the whole determines the formation of its own components. This may be expressed in other terms: the entity is already capable of realizing different potentialities, but they are not yet localized in a mosaic of pieces. The mosaic assumes a physical form only later and under the influence of factors that are either unknown or only beginning to be recognized.

It is not infringing upon the laws of determinism to think that an entity may resolve itself into distinct components, each with its own individuality. Such difficulties were not foreseen by philosophers like Kant, Lachelier, and others. We can avoid difficulty by expunging the term *teleonomy* and substituting the term *potential.* Among biologists, the concept of teleonomy is far from being universally accepted; from a pragmatic point of view, claiming a teleonomic solution is a lazy man's explanation. But there is a moment when one can no longer avoid considering such a solution. Many biologists will object to such a concept, but what better solution can they propose? That all nature develops according to rigid and inescapable mechanical processes? That all future generations are infallibly foreseen in the egg of a single generation and, indeed, in all the eggs that preceded it? Everything would then have been determined since the emergence of life on Earth, ever since the first living organism, whatever it may have been. There would not have been contingency, liberty, or innovation in any living organism — in its physical realization or in its psychological personality. One is forced into a strict materialism.

It is difficult to suppose, a priori, choice among possibilities, liberties, and contingencies, but it is as difficult to imagine that we and other living organisms are merely machines and that psychic life is only an unimportant efflorescence of living substance. We arrive at a problem that many philosophers have considered without being able to resolve. Nevertheless, teleonomy exists; one cannot escape it. First-rate biologists, like Lucien Cuénot, Albert Vandel, and Pierre P. Grassé, have asserted it. This present sharing of the idea among scientific colleagues is not an argument in its favor but does demonstrate that the question is unavoidable.

The ideas developed here may satisfy professional philosophers, who may take them to be self-evident, while certain biologists may be shocked by them, because of a natural distrust of finalism. They may prefer the word *teleonomy* to the word *teleology*, because the latter includes purpose. But terminology is not important. Facts are important, not theories or fragile hypotheses. However, we cannot disguise the difficulties of explaining the processes of life.

V.2. The Double Helix and Messenger RNA

Molecular biology has made remarkable progress since the midcentury. Among its numerous discoveries, two have acquired considerable significance and public recognition: DNA and messenger RNA. Together, they launched a movement that continues to stimulate the imagination of biologists. It is difficult to give a short account of these two major discoveries, and there is no point in rehearsing here the main lines of these ideas. We shall merely try to extract their meaning and relate it to the ideas of causality and teleonomy.

We know that the genetic traits of an individual are transmitted by the individual's DNA (deoxyribonucleic acid); it is to be found in the nucleus of all of the individual's cells. Two questions may be formulated. The first is, How does it occur that, although from the beginning all the organism's cells contain all the future individual's genetic material, a differentiation occurs among them to permit their development into the specialized cells of its organs; and why are certain cells destined to become one organ rather than another, a brain rather than a kidney, an ovary rather than a femur? The second question is, How does the DNA of the nucleus transmit information to the cytoplasm in order for the latter to synthesize the various proteins, which are, in effect, the definitive and true expressions of differentiation? There are numerous other questions that derive from these two, many of which, despite some claims, are not yet answered.

It is to the credit of a pair of well-known biologists, James Watson and Francis Crick (1953), that the structure of DNA was unraveled in 1953. DNA is constituted by two intertwined strands—the double helix—of chains of sugar phosphate molecules. These strands are bridged by purine bases (adenine and guanine) or pyrimidic bases (cytosine and thymine). These pairs of bases are complementary because they are united within the DNA by hydrogen bonds. Normally, adenine (A) is tied to thymine (T) by two hydrogen bonds, and guanine (G) is tied to cytosine (C) by three hydrogen bonds. Between the strands of sugar phosphates, these bases are arranged like the rungs of a ladder.

Since all the cells of a multicellular organism contain the same molecules of DNA, they should all be able to synthesize the same proteins. In fact, however, each type of cell synthesizes only a certain well-defined number of all the proteins coded in the cell's DNA. Within the cells of eucaryotic organisms (those that possess a well-defined nucleus), the sequence of nucleotides in the genes is transcribed in the DNA of the nucleus.[4] A *gene* is defined as the segment of DNA that codes for a given protein.[5]

The RNA leaves the nucleus and passes into the cytoplasm, where it is translated into protein. There are thus two primordial phenomena that intervene between the presence of DNA in a nucleus and the formation of RNA in cytoplasm. One is a transcription of DNA into RNA. The other is a translation of RNA molecules into protein. Once the DNA has been transcribed, the RNA is transmitted to the cytoplasm. This process was discovered by biologists Jacques Monod and François Jacob, who called the resulting product "messenger RNA." Thus, eucaryotic cells have many possibilities for development. Regulation may affect the choice of the DNA segment to be transcribed, the sequences to be transcribed onto the messenger, and the rhythm of translation of RNA molecules into protein.

Toward the end of the 1970s it was already known that several operations were necessary to modify primary RNA, which is directly transcribed from DNA onto "mature" messenger RNA. The latter may be transcribed into protein by means of several modifications. In effect, different structures are added to the two extremities of the RNA chain, certain of the chain's nucleotides undergo chemical changes, and what is still more surprising, the primary "transcript" can be cut and rearranged in several fashions. It thus produces different RNA messengers and, through them, different proteins. The result is thus, in some sense, a production of a contraction of the DNA molecule, which eliminates several segments, called *introns*. The remaining segments, called *exons*, are the only ones that remain active in the coding of RNA.

The first operation, the transcription of the DNA onto the RNA, is performed by an enzyme called RNA polymerase. This enzyme binds to the DNA at a particular site, indicating the beginning of the information coding for the RNA. It thus chooses the first nucleotide of the RNA chain. The enzyme then moves rapidly along the DNA molecule and adds to the first nucleotide the other nucleotides, which then constitute the RNA molecule.

> When the RNA polymerase transcribes a region of a strand of DNA, the nucleotide that it adds to the RNA is the complementary nucleotide to the one found on the DNA. Because of the molecular structure of the nucleotides, if one type of nucleotide is given, it can, in effect, establish hydrogen bonds with only one of the other three. Thus a C and a G can establish hydrogen bonds only between each other, just as an A and a U (or a T) can establish hydrogen bonds only with each other. It is the hydrogen bonds that permit the complementary strands of DNA to come together to form a double helix. The RNA that is transcribed is consequently a complement of the fragment of DNA from which it was formed. Since

> the RNA contains information about the part of the genome — the set of all genes within a cell — that was transcribed, it in fact constitutes a matrix for the chain of amino acids that constitute a protein. The enzymes that "decipher" the information coded in the RNA "read" the nucleotides in groups of threes. Each triplet of nucleotides, which codes for one of the twenty amino acids of proteins, is called a *codon.* During the translation, the reading of a codon produces an addition to the protein chain being formed with the amino acid corresponding to this codon. (Adapted from Darnell, 1983, 36)

The combinations that can be formed from a constant number of molecules is astonishing. But one might ask, How many great works of literature have been created with an alphabet of only twenty-six letters? Thus when our mind or body creates something novel, it is by virtue of the composition or the arrangement of constant and predictable molecules.

But with DNA molecules, one can create millions of combinations. Furthermore, every day something unexpected contributes to creating a new substance — for example, *mature* RNA. Messenger RNA, such as is formed by DNA, is not mature; it is transformed into active messenger RNA by the elimination of introns and contraction of the linear molecule. Despite its complexity, such elementary and monotonous grammar may appear simplistic when compared with the highly complex and apparently original phenomena of life. Why should RNA be active only when introns are eliminated? Such as it is, the code often reveals new phenomena. How and why is the inception of the synthesis of RNA accomplished? Molecular biologists have an answer for everything, and their solutions are generally acceptable. Let us consider an example:

> Ronald Evans, Michael Rosenfeld, et al. . . . have found in the genome of a rat a transcription unit with two poly (A) sites. This unit carries the sequence that codes for two proteins, calcitonine, a hormone synthesized in the thyroid gland, and a recently discovered neuropeptide, a hormone of the nervous system synthesized in the hypophysis. . . . Thanks to studies undertaken since 1975, it has been proved that the genes of eucaryotic cells may be regulated by the "differential" maturing of the RNA. However, it is not yet clear what may be the importance of this tool within the collection of regulatory genetic mechanisms. (Ibid.)

It is clear that, despite considerable progress in the study of the mechanisms of life, much remains that is mysterious. Everything may be definitively resolved into mechanistic processes, but what triggers them? There will always remain enough that is inexplicable to leave space for contingency.

Isn't the complexity of the organization of organic but nonliving matter remarkable, matter that is secreted, structured by living matter, capable of transcribing and transmitting messages in a coded language? We realize these are symbols, figurative images of an inert reality. And when one says that RNA polymerase passes from nucleotide to nucleotide of the DNA molecule, that too is only a way of speaking. RNA polymerase does not really move; it is there, at work, but it reacts in succession with the nucleotides, which are arrayed in a linear sequence.

There is, therefore, a certain order in the operations, in the reactions, and that order is programmed—we cannot avoid using this word—is established in advance. It is determined, but not quite immutably determined, because the enzyme can choose the site at which it situates itself, the codon, the segment of the chain with which it will react. The word *choose* is probably misleading, because all these reactions take place according to a mechanism that is probably inescapable. However, these mechanisms have a certain lability that permits the transcription to begin at a variable site, which may also be determined—possibly even strictly determined. But how is the inception of this process triggered? It is at this point in the argument that the explanations become uncertain and stumble against chains of causes that succeed each other endlessly. It may be there that contingency appears. It is also enough to represent different modes of mending the RNA in order to imagine that everything is not always strictly and definitively fixed in advance.

It is difficult not to invoke the laws of strict determinism (which is compatible with contingency through the special conditions of realization [I.5.5] [VII.2]), and those who try to avoid it risk considerable embarrassment. But there are so many possible techniques of regulation of the formation and regeneration of RNA that one need not abandon hope of finding contingency in its workings.

Notes

1. The normal fibula of a bird is constituted of a small, slender rod that terminates in a point at the level of the middle of the tibia.

2. This incertitude is due in part to the fact that one can reason only from skeletal evidence. If one knew more about the soft tissues of prehistoric animals, it might be easier to decide.

3. Embryos with a single eye in the middle of their forehead are called cyclops, like the Cyclops of the *Odyssey.* There is a continuous scale for *cyclocephali* between the normal and the cyclops. Symmeli are monsters with two limbs fused together. There are also monsters intermediate between the normal and the symmeli. Anencephali are embryos whose initial cerebral plate is not closed into a cerebral vesicle. The result is that their brain does not develop.

4. Eucaryotic organisms include all animal and vegetable species except for bacteria, viruses, and cyanophyceae (blue algae), which are called *procaryotes.*

5. The expression *coded for*, although incorrect, is generally used to mean containing the information regarding a certain substance or synthesizing a certain substance.

VI

Economics, Turbulence, and Chaos

To the best of our knowledge, the idea of bifurcation has not been used by those researchers in economics who are not mathematicians. Even in physics, explicit applications of this concept have been only fairly recent, despite the fact that Henri Poincaré proposed it in 1885, following suggestions in the work of physicists Sir William Thomson and Peter Guthrie Tait. The word *bifurcation* may evoke an image of railroad track switches, where several directions are possible. It is a question of qualitative changes along a course, changes that may be associated with quantitative changes.

At the present, we are concerned with the concept of turbulence in economics. But in the background are precisely the same notions of singularity and bifurcation as in physics. When many successive bifurcations appear, expressing the multiplicity of possible solutions — and especially when they are in spatial and temporal proximity — one can see behavior that in the long run leads to chaos and turbulence. In the near future, considerations of qualitative changes will play a greater role in the understanding of economic systems. It is not difficult to find examples in these days of economic disquiet. The pages that follow give concrete examples of transformations in contemporary economic structures.

Economics can also benefit from the type of reflection that has cast physics problems in a new perspective. In order to become a true science, economics has tried since its beginnings to model itself after Newtonian, mechanistic concepts. Why should economics not try to accomplish what physics has accomplished — that is, marshaling mathematical theories and recent concepts to address problems stemming from qualitative structural changes? It goes without saying that the problems of homo sapiens are of a different order than the problems of physical objects, but nevertheless the objective is worth the effort. We shall see its implications.

The phenomena of bifurcation are in part related to variations in previous and succeeding data. They depend for the most part on variations in structural parameters, and they govern the changes that evolution brings about.

Present-day economic turbulence is our starting point. We ask whether these perturbations are related to randomness and chaos, whether they are subject to determinism, and whether they are capable of leading to some new order.

We must immediately realize that we cannot yet speak about a true science of economics, although one often does, as though it were a science that could permit us to overcome the difficulties that beset us. Economic reality is, in this sense, different from physical reality and even from biology. Although economic reality may be analyzed precisely, one is struck by how many points of view there are among economists. There is too much interpretation, most of it subjective. Economic facts are in common view—to a greater extent than physical facts—and everybody thinks that he or she is qualified to understand them.

Of course, economists have received a thorough education that helps them avoid simplistic and hasty opinions. But there are many schools of thought, and even if one considers oneself independent, one still belongs to one or another of these schools. So, although economic reality is unique, it does not have the distinctness of physical phenomena. The greatest difficulty lies in the fact that, while one has exact descriptions of elementary phenomena in the so-called exact sciences and one may apply to them combinatorial theory, economics has not yet produced the concept of an elementary phenomenon. And is such a concept even possible?

The founders of economics believed such a concept was indeed possible: the nineteenth century was characterized by a continuing attempt to isolate elementary phenomena, from which everything else would follow. But today's world is so complicated that we have had to give up that ambition. That is why the idea of complexity may be so appropriately applied to economics and why interdisciplinary contributions may be so useful. The founders of economics had to simplify their working hypotheses without always recognizing that they were in fact working with hypotheses. What they took to be the ultimate foundation was pure and perfect competition in an imaginary market, and thus it was elementary. Without being fully conscious of it, they thought they had isolated the elementary phenomenon, thus imitating the astronomer who claimed that the phenomena exhibited by the stars, despite their apparent complexity, were simple phenomena because there were no strong interactions among them. Astronomers were the first to succeed in celestial mechanics because mathematics permitted them to formalize their observations of nature in the simplest of cases.

It was natural that economists should want to imitate astronomers. But the complexity of the economic domain got the better of them. Their model of the economy has remained mostly intellectual. We continue to use it, but given the current state of knowledge, it is inadequate to explain economic reality and insufficient to guide actions. Nevertheless, since we still use that model as a point of reference, we may wonder whether the concept of turbulence could not improve it.

Nineteenth-century economists considered a crisis an exception: it contradicted the order the mind imagined. And yet, since crises recurred periodically, they could no longer be regarded as exceptions. Economists then tried to integrate the idea of crisis into their model, thus producing a first understanding of turbulence as an apparent disorder that could lead to a recovery of the initial order. A new economic science was erected upon this idea. But the world wars intervened, and the mechanisms created with such rigor, using mathematical tools, were no longer suited to the analysis of economic realities. The idea of turbulence did not disappear. To the contrary, it surfaced in economics in another form, *structure* and *conjuncture.*

For mathematicians and physicists, the structure of a phenomenon appears when it emerges from numerous experiments that show that some substratum remains unchanged beneath certain apparent changes. This conserved substratum expresses what is essential to the phenomenon. Economists have a different interpretation; the concrete aspects of their structure are more existential than essential. Structure is found in all observed reality and is thus a form of architecture characterized by the reciprocal proportions of the elements of a set. The structure of a population is expressed by a pyramid. For example, the working population has a structure containing three categories: agricultural and extractive, industrial production, and service. Foreign trade may be described by the ratio of imported to exported elements: foodstuffs, raw materials for industry, and manufactured goods. Families also have a structure, as does capital.

When economists contrast crises of conjuncture and crises of structure, they tend to think that in a given structure of society events succeed each other in a given way with certain characteristics. This structure is a result of human action, a heritage of the past, of history. For instance, in the structure called *capitalism,* we can see a certain form of conjuncture — variables conjugate, encounter. There are fortunate encounters, like prosperity, and unfortunate encounters, like recession. But for a certain school of economics, if we change the structure of society, the same conjunctures will not occur. To take the argument further, there are no unfavorable conjunctures. Taken to its logical conclusion, there are no more conjunctures of any kind. This differs from the natural sciences, where human power cannot transform structure and, hence, conjuncture.

We return here to our discussion of determinism. Things are determined within a given structure, and if the structure changes, these same things may no longer be determined. A determinism of other things will appear. Turbulence reappears. Even though a situation may be structurally stable within a certain spatiotemporal domain, turbulence may occur if that domain is small, if its duration is short, and if it is the anticipation or manifestation of a structural change.

What we propose here does not correspond to accepted opinion among economists, most of whom believe that, whatever the political regime, relations exist that cannot be other than what they are. They treat these relations as though they were actual laws and were thus constant and ubiquitous, like the laws of gravity and expansion of material.

Laws are expressed in relation to the concept of equilibrium. The science of economics is founded upon it. One may ask, however, whether the concept of equilibrium is sufficiently broad to support a science of things in perpetual motion. The concept of equilibrium expresses, in effect, the idea of remaining the same. The idea of stability is even more suggestive.

There are two types of stability, the stability of position and the more profound stability of structure. After stability of position is achieved, if a perturbation occurs, whether it originates within the system (endogenous) or outside of it (exogenous), whether it is determined or erratic, the question arises, Will it be damped by its own action? The perturbation begins, quite obviously, by disturbing the equilibrium of the system. But if one lets nature work without interference, the disequilibrium may disappear, in which case there is a return to the initial equilibrium.

But if it is a perturbation that persists (turbulence), it may prevent the previous order from reasserting itself. This is true of the current economic situation, which is characterized by both unemployment and inflation. Nonetheless, we must consider equilibrium, or stability, in several ways: if turbulence prevents a return to the former order, would it not be possible to consider it a harbinger of a new order, not only different from, but possibly even better than, the old? In the jargon that economists prefer, *turbulence may induce a change in structure, which transforms conjuncture.*

In fact, this is not how it was first explained. To avoid upsetting the interpretation by equilibrium, chance was invoked; but chance is the equivalent of a certain disorder, a Till Eulenspiegel, a demiurge.[1] It would be curious in this context to consider the role played in contemporary economic models by variables called *stochastic*, or *random.*[2] These variables permit us to retain the determinism of ad hoc equations that lead to rigorous solutions. Since reality almost never satisfies these equations, a margin of random effect saves their validity, but within limits.

But there is another means of constructing a model. Until recently, the models used by economists have been linear, that is, based on linear equations with or without a delaying term. We now know that these models are too crude to be meaningful, and modern mathematicians have constructed new models, such as Yves Balasko's (1978) for the economic equilibrium and catastrophe theory and Daniel Royer and Gilbert Ritschard's (1985) for the application of qualitative methods of structural analysis.

What interests us here is the extent to which turbulence can be introduced into the model to render it closer to reality. We no longer start from a representation supposedly perfect but, rather, from turbulence (which is not really well named; our discussion of chaos in I.7 seems more meaningful). Beyond a certain threshold, turbulence or chaos may lead to disaster, to a fall into a chasm that the etymology of the word *chaos* suggests. On the contrary, moderate or limited turbulence may perform a regulating function in economics just as it does in fluid mechanics and intramolecular motions.[3] It should be determined, therefore, whether turbulence in economics obeys certain laws and whether, retaining the vocabulary of chaos, it has states of deterministic chaos that render a new possibility. (Of course, this way of looking at things could lead to language reform, which is particularly difficult because linguistic habits are thoroughly ingrained.)

It may be appropriate here to analyze the related ideas of *complexity* and *teleonomy*, which have stimulated this discussion. Complexity is an obstacle to scientific theory. As already shown, complexity has been effectively eliminated by primary theories; we have tended to think in pieces. Such simplicity is, however, dangerous and may lead to absurdities. Of the two principles that emerge from Jules Lachelier's argument (1924), only one was thought pertinent to economics, the one that asserts that phenomena form series for which the existence of any one element determines that of its successors. But economists are now beginning to recognize the second, which asserts that these series form, each in turn, systems in which *the existence of the whole determines the existence of its parts* (see I.5 and V.1).

This is one way of recognizing the complexity of things. Taking one isolated variable to explain another, equally isolated variable is what scientists have often done. A famous example, still frequently cited, is the quantitative theory of money, which claims that the quantity of money in circulation explains the general level of prices. Among other things, there ought to be agreement about what is meant by "the quantity of money" and how "the general level of prices" is calculated, because it is a concept rather than an observable phenomenon. But we need not discuss that here.

Once a variable is extracted from its context, it is not possible not to have changed that context. It is like a castle built of playing cards, where moving one card may bring the castle tumbling down. Scientific method must therefore seeks ways of studying a phenomenon while disturbing it as little as possible.

Complexity is inseparable from teleonomy. When considering a spatial entirety, one must also consider that entirety in time. It is the end result that determines present behavior. The word *end* emphasizes teleonomic cause; the idea of *end* is fundamental, since the word has two meanings, that of *completion* and that of a *goal* that one attempts to realize.

How can the concepts of biology be used by economists? The terms *final cause* and *teleonomy* do not have a good reputation among orthodox scientists. According to them, it is science only if what they call *causality* is the exclusive relation among phenomena, and it is therefore more appropriate to speak about discovering *relations among phenomena.* In any event, it is not a question of renouncing the attempt to determine relations among phenomena, but such research alone is not sufficient. The present state of the world is ample proof of that: analyses of unemployment, inflation, and economic growth have proven their inadequacy. And this is not the place to discuss the difficulty of understanding the phenomenon of inflation, although we might consider viewing inflation as a manifestation of entropy. Passing from one state to another expends minimal entropy. The objective in dealing with the current price increases is to avoid wasting entropy in order to avoid introducing unnecessary inflation.

There is no shortage of theories that claim to be exclusively relational. New ones are proposed every day. Do they furnish solutions? There is reason to doubt it. Can we avoid the difficulty, as has been done in biology, by replacing the term *teleonomy* with the term *potentiality*? It is not obviously possible.

What direction is economics taking? Must it restrict itself to constructing curves on the basis of past experience in order to extend them into the future? Is the future indeed exclusively an extension of the curves of past experience? Obviously, we must not disguise a major problem. Who will define an *entirety*? Who can define it? Are there not as many definitions as there are authorities? This objection is not devastating. One may identify several objectives and, for each of them, seek the potentials implicit in the economic decisions. One solution or another will appear as a function of these objectives.

In any event, although it is impossible to define a comprehensive entirety, the entirety of the end of time, we can distinguish various temporary frontiers within total time. The distinctions that Jean Fourastié (1979) prefers may lead to proposing a short-term entirety, a midterm entirety, and a long-term entirety. One can therefore draw up a table of potentialities according to diverse hypotheses. We recover here the idea of different scales of time, the point of departure for this volume (see I.2). Nonetheless, for a given instant, and for that very instant, there cannot be, simultaneously, several objectives.

In current discussions of economic growth, decisions regarding investment are considered expressions of causality, because they are made in the light of prevailing unemployment or inflation. And we might think that it is by the discovery of such causes that science is constituted, but the results have not been very conclusive. A physicist might assure us that phenomena can be interpreted causally when they are explained by forces and teleonomically when we want to act upon reality. What is more teleonomic than Fermat's optimization of time as a functional criterion? (IV.1). Economists have been particularly partial to optimization procedures since Vilfredo Pareto's theory was published (1896), but did they then understand it as an expression of teleonomy?

One of the most classic, even banal, ways of looking at science is to distinguish two objectives, the acquisition of knowledge in order to understand reality and the undertaking of actions in order to modify reality. Different methodologies correspond to these two ways of looking at science: short-term interventions seek to resolve local problems, but understanding encompasses the universe. Understanding requires us to think globally, even though we must still act locally. Aristotle's efficient causes, which John Stuart Mill regarded as anterior causes, seem related to *local* causality. Are Aristotle's final causes thereby related to *global* causality or the diverse manifestations of teleonomy (see I.5)?

Of course, the concepts of cause and effect may appear naive, deriving from an anthropocentric view of the world. When the interval between the inception and the conclusion of an enterprise is progressively reduced, how can one distinguish between what is cause and what is ultimate effect? We may invoke Wassily Leontief's argument about whether "history may be written backwards" (1974, 53). If one works backward from some ultimate act toward a determination of what happened at some initial epoch, it is simply another procedure adopted so that causality might not be opposed to teleonomy.

In any event, how can economists now avoid taking these conclusions into account? We recognize this necessity in the mishaps that result from a purely quantitative measurement of the economy. At its inception, economics was oriented toward politics in the most noble sense of the word, but in the nineteenth century it sought to become a purely cognitive discipline.

Now, toward the end of the twentieth century, economics is rediscovering that it must dedicate itself to action, since cognition and action influence each other.

The science of economics has been created gradually and has tried to emulate the concepts and methods of the so-called exact sciences, but economic actors are aware that they are being observed and, furthermore, are endowed with freedom, which is not true of the subjects of the exact sciences. Therefore, economics by its nature cannot conform rigorously to the models of the exact sciences (Kline, 1987).

For it to be otherwise, we would have to assume a natural order of economics independent of human volition. This is exactly what was supposed by the "physiocrats" of the eighteenth century, who are rightly considered the founders of the science of economics. They thought only human error and miscalculation prevented that natural order from being realized. This is close to the way ecologists argue these days: that the pollution, destruction, and exhaustion of natural resources resulting from the "progress" of industrial civilization have degraded an otherwise natural order.

However, the word *nature* has at least two meanings, *existence* and *norm*. Nature is the set of phenomena in time and space that the professional natural scientist recognizes. Nature is also whatever is in conformity with the essence of a being, in particular, a free and reasonable human being, and whatever determines that being's vocation. It asserts what ought to be, in normative language.

The first meaning is that of the physiocrats. Their name comes from φυσις, which shows that economics is related to physics, in some sense. But even in its first sense, nature appears to be organized in some teleonomic scheme. In other words, to the extent to which the human being is integrated into nature, in particular in hunting and gathering societies, nature regularizes itself. Like the lives of plants and animals, the lives of humans fit into nature's scheme. Biology has shown us that such regularization in fact occurs.

But humans have tried to dominate nature, and to this extent the ecological natural order has not been respected. It is thus difficult to imagine a teleonomy of ecological origin or an uneconomical ecology. The word *economics* is not comparable to the word *ecology*. The stem, *nom*, from the Greek νομος, implies the idea of an order to be realized and not merely to be discovered: automorphism is not enough to produce the order one desires. Furthermore, humans sometimes risk producing the contrary of what they desire and one can analyze the more perverse effects of these results on the economy.

Finally, the ideas of complexity and teleonomy seem likely to renew the science of economics. Having sought to imitate the laws of physics by discovering the laws (*nomics*) of human domestic (*eco*) organization, economists might ask whether the discipline has not deviated from its primordial vocation. Certainly, some economists have established theorems that advance both descriptive and mathematical economics, but they no longer accord the suffix *nomics* the meaning it had when economics was created: the function of participating in an order being created. Theorems should serve the construction of that order, but they themselves do not constitute that order, which indeed is constantly being destroyed and recreated — and which we continue to seek.

Notes

1. Chance cannot explain the coincidence of numerous causal chains (see I.8). Further, data show that nonzero correlations among properties of a system prove that the system is not disordered (see I.6).

2. The word *random* carries no philosophical meaning regarding questions of chance and disorder (see I.8).

3. We have seen in chapter III that turbulence in the atmosphere is a regulating mechanism for Earth's climate.

VII
Conclusions

VII.1. General Remarks

In ancient Greece, a philosopher was also a scientist. Pythagoras and Democritus introduced mathematical models of physical systems. The Eleatic philosophers and, subsequently, Socrates and Plato established noncontradiction as a law of thinking: language bore witness to order and organization, dialogue discovered ideas. Aristotle established logic, physics, and metaphysics and brought science down to the realm of perceptible phenomena. Most of the concepts and principles were asserted fairly explicitly, and philosophy was highly developed.

In the seventeenth century, scientists like Galileo, Descartes, Pascal, and Leibniz were also philosophers. But since Newton, who developed both the elements of classic mechanics and the method of the experimental verification of theory, the very rapid advance of science led to its specialization into disciplines and then to its separation into philosophy and science—which, however, should not have, but in fact has, obviated exchanges of views.

Although among our contemporaries, some masters like Henri Poincaré have achieved rather broad views encompassing both science and philosophy, it has become increasingly necessary for scholars to work in interdisciplinary groups and colloquia to devise syntheses of science and philosophy.

In the spirit of Descartes, we can say, "I think, therefore I am." We think that we think and become thereby aware of our personality and free will, coming to recognize their distinction from the exterior world, of which we have obtained a partial acquaintance by means of our senses and the interpretations formed from their reports. These interpretations are formulated in the context of our mental structures and the concepts, principles, and methods inherited from our predecessors. The concepts of time, space, and motion are implicit in Heraclites' maxim, "Everything flows." The intelligibility of phenomena is a basic principle of thought, together with order, law, causality, teleonomy, and rationality. The concepts of criticism, experimental verification, and determinism constitute the significant intellectual progress achieved by modern science.

Both the accumulation of information and the increase in scientific knowledge of the observed universe comfort us with the idea that our minds may achieve an understanding of our own condition and the workings of the world about us. However, at the same time, we recognize our limits, because, on the one hand, this process converges, permitting us to class phenomena under laws common to them all and, on the other hand, it diverges by constantly posing new and more general questions, thus highlighting the inability of science to attain complete knowledge of ultimate reality.

Introspection is natural to us, as is curiosity about the world. But despite our efforts to attain scientific and philosophical objectivity, we do not succeed in disengaging ourselves completely from anthropocentric views, except possibly in mathematical abstractions.

It might be useful to seek still more general ideas about places and kinds of life different from ours, about universes that are multitemporal, or even transtemporal, and spatial, but that would be speculative. We consider here only the universe we perceive through our senses; and we apply to this universe the usual methods of systematic investigation, using the accumulated information and modes of thought of the several disciplines of the authors.

The variety of methods, concepts, and languages of these disciplines made the authors attempt (in chapter I) to reduce the ambiguities and confusions of language that might have otherwise led to fundamental errors of interpretation. Aspects of phenomena, and also the expressions of the laws that describe them, depend not only upon their intrinsic properties but also upon the means and scales of the spatiotemporal observations used. The latter must always be chosen appropriately for the problem at hand. The fact that different aspects of phenomena can be observed at different scales is not logically or scientifically contradictory, but complementary.

VII.2. The Interpretation of Phenomena through Causality and Determinism

Determinism, considered as an application of causality in the sciences, may be expressed in diverse modes and with diverse degrees of precision.

The language and the methods of mathematics help unmask inconsistencies, confusion, and contradictions, and they permit us to arrive at authentic abstraction when a problem can be formulated within their terms, most notably in the physical sciences. The rules of physicomathematical determinism may then be expressed precisely, as may relations with necessity, contingency, freedom, and even unpredictability (see I.5). These rules include two general conditions:

1. The physical system being considered is defined as a set of inanimate bodies, circumscribed in space and time; its state in the spatiotemporal scale elected is represented by a finite number of variables, the unknowns.
2. All the changes that this system may undergo are governed by principles and laws expressed by equations, whose number is at least equal to that of the unknowns.

These general conditions are *necessary* because this system cannot be other than what it is and because it cannot obey principles other than those that govern it, principles like the conservation of matter, energy, and variation of momentum. Thus, in the case of a fluid, since the general equations represent all its possible states and motions, its behavior necessarily conforms to the laws of physics.

Next, the rules that describe the special conditions of the system being studied include

1. the constraints imposed by its boundaries and by the free surfaces at the limits of its spatiotemporal domain,
2. the forces, such as gravity, electromagnetic forces, and radiation, that are exterior to the system but that act upon it, and
3. the initial conditions or state of the system at an arbitrary moment, taking into account previous states.

These special conditions are *contingent* because the principles of physics are variational and thus formalizable as partial differential equations whose integration introduces arbitrary constants and functions that correspond to the circumstances that describe the system.

Boundaries and free surfaces may have alternate forms, may or may not be permeable to matter and energy, and may or may not change with time. Similarly, exterior forces may or may not act upon the system, according to their modalities and variable intensities. For example, water confined behind a dam with several outlet channels may flow through one or another of them. The choice of outlet is not necessary in terms of the physics of the confined water but contingent, depending upon the free choice of the person who adjusts the gates, which may also be controlled from a distance by actions equally contingent and even more freely chosen. The initial state of the water corresponds to immobility or a particular motion, according to whether the gates are closed or open.

If all the general and special conditions for the description of the particular circumstances are satisfied, the subsequent states of the system, within its spatiotemporal domain and at the chosen scale of observation, are completely determined, even when they are complex and turbulent (see I.5 and II.2).

Physicomathematical determinism, which is the strictest form of determinism, is compatible not only with the necessity imposed by the laws of physics but also with the contingency inherent in the particular circumstances of the situation under consideration. Freedom may be exercised within the context of that contingency. A complex phenomenon that cannot be predicted exactly in the long term by the techniques now available is nevertheless not necessarily incompatible with the most rigorous determinacy by the principles of physics (see I.5.6).

Physical determinism may also be discovered through experiments that show the replicability of a phenomenon to each experiment performed within the same system, under the same conditions, within the same boundaries, subject to the same external forces, and with the same internal conditions. This type of determinism is also compatible with necessity, contingency, and freedom. And statistical determinism—which is based upon the replicability of average properties under the same average conditions for each series of experiments—is not contrary to necessity, contingency, or freedom.

Mainly in the physical sciences, and to a lesser extent in the biological and social sciences (Kline, 1987), determinism is expressed according to the languages used and within the corresponding data, which are never perfect. This method ought to be used in all the domains where scientific knowledge permits a satisfaction of the above conditions, even if the phenomena under study are complex and turbulent; but it is not applicable when scientific information is insufficient. Philosophical consequences must be drawn cautiously. As we shall see shortly, such has not always been the case—for instance in the contrary examples of Laplacian universal absolute determinism and of Democritean and Epicurean "chance" and "disorder."

The conditions for physicomathematical determinism are fully satisfied for well-defined physical systems depending upon the laws and principles of classic Newtonian mechanics applied at the human scale. For more than two centuries, these principles have been experimentally verified with remarkable precision. Classic mechanics is thus subsumed within rigorous physicomathematical determinism, particularly in the case of turbulent fluids (see I.5.7 and II.2). Einstein's theories of special relativity and gravity generalized classic mechanics to phenomena whose velocities are not negligible with respect to that of light, and they too are compatible with physico-mathematical determinism (see I.5.7 and IV.2).

Statistical mechanics, which studies the states of matter at a molecular scale, is based upon the same laws and principles as deterministic classic mechanics. Because the phenomena studied are complex, due to the very short time scale of molecular motions, which is not susceptible to detailed measurement, it is customary to treat them statistically. But this does not imply that their behavior is not determined. The above methods yield a statistical determinism for average quantities, which is to say for their properties on a human scale (see I.5.7 and IV.3).

At the beginning of the twentieth century, the study of the behavior of matter at a subatomic scale, which had until then been treated by quantum theory, led Heisenberg to posit the uncertainty principle, according to which the position and the velocity of a particle could not be measured simultaneously because they are inextricably related. From a philosophical point of view, this principle was thought to cast doubt, if not upon causality, upon determinism within that microscale of observation. However, the sense that quantum mechanics attributed to the word *particle* has no relation to our intuition of a small fragment of matter or quantity of energy. The *physical elements* we think about are really *fields*, which are more or less modulated in space-time. The modes of thought that localized photons or split an electromagnetic field into two or several photons have proven completely inadequate. States of matter are actually described by statistical functions, whose changes are fully determined (see I.5.7 and IV.4).

In the nineteenth century, Pierre-Simon Laplace imagined applying determinism to the universe, whose present state would be the result of its past state and the cause for its subsequent states. For an intelligence that recognizes all the forces of nature and all the elements that compose it, nothing would be uncertain. Laplace took care to emphasize that the human mind is no more than a pale draft of such an intelligence, which must always remain infinitely remote.

Now, according to one philosophical doctrine, everything in the universe—particularly human actions—is connected in such a fashion that, *things being what they are at a given moment, there is only one state compatible with that initial state for any other moment, prior or subsequent.* But since a complete definition of the universe is not given, all the laws and constraints that govern it, including those that govern human behavior, are not known and cannot be formulated in closed equations. This doctrine, which would eliminate contingency and the freedom to think and act, is the result of an ill-considered extrapolation of determinism to a situation where its conditions are far from being realized because of lack of information (see I.5.8).

In recent years, studies have been made of complex and fluctuating phenomena, which are in fact the rule rather than the exception in nature. A complete description of such phenomena would require so much information that there is a temptation to think that they are unintelligible and subject to disorder and chance. On the contrary, however, at certain levels of observation, structures appear that reveal intelligible forms of order. Such structures have sometimes been interpreted as illustrating the principle of "order produced by disorder" attributed to Epicurus. But an a priori renunciation of intelligibility does not seem justified; disorder may be only apparent and may result only from our temporary inability to discover the order in complex situations. So disorder may indeed be subjective; otherwise, one would have to define the disorder of a system objectively as an absence of relations among the elements of the system and their properties or, in the language of probability theory, as their *independence*, which implies that all correlations are nought, and the result would be the antithesis of determinism. Thus it is sufficient to recognize the existence of nonnegligible correlations to reject the hypothesis of disorder (see I.6 and I.7), which in addition must obviously also be rejected a fortiori when the phenomenon is in fact determined (see I.5).

As for *chance*, defined as induced by the coincidence of phenomena that belong to series that are independent with regard to causality, since the probability of such coincidence is then equal to the product of the probabilities of each series, it tends toward zero rapidly when the number of series increases. So the occurrence of events by reason of a coincidence of numerous phenomena cannot be explained by chance and shows that these phenomena are not independent.

Ilia Prigogine studied the thermodynamics of unstable, fluctuating states and discovered consistent, hierarchized, evolving, and dissipative structures emerging from the complex fluctuations of chemical reactions (Glansdorff and Prigogine, 1971; Prigogine and Stengers, 1979). Extending these ideas to other domains, such as turbulence, physics, the sociology of insects, and economics, he formulated a principle of "order produced by fluctuation" (Prigogine and Stengers, 1979). This assertion should have helped us avoid the common confusion between the meanings of the terms *fluctuation* and *disorder.* There are in fact fluctuations, even very complicated fluctuations, that are completely determined and thus *not at all disordered.* As we have seen, the theory of dynamic systems represented by closed differential equations offers examples of such complex fluctuating phenomena, which have been called chaotic but which are, nevertheless, strictly determined (see I.3 and I.7).

Turbulence in fluid flows, which is a general phenomenon, provides a particularly meaningful example of the term *deterministic chaos*, which we distinguish from *disordered chaos* by spatiotemporal correlations (see I.7). It is one of the most complex fluctuating physical phenomena at the scale of human observation for which the fundamental laws and principles of classic mechanics and physics may still be expressed by a number of equations equal to the number of unknowns. Were the detailed state of the system at a given instant described, the boundary conditions stated, and the exterior forces acting upon the system completely accounted for, one would be able to show that the state was physically and mathematically determined. Furthermore, measurements of turbulent flows have shown high correlations among them. These also argue against disorder (see I.5, I.6, II.1, and figure 6).

When these dissipative flows are observed at different scales, changing but coherent eddy structures may occasionally appear. They become the locus of interactions and break down into increasingly small structures or are transformed by regrouping into larger structures. These new forms of order proceed from prior forms of order and not from disorder or chance. Thus *order* may be *complex* and *fluctuating*; it may *evolve* through changes in *scale* and *structure.*

The study of the behavior of the atmosphere and hydrosphere, described in chapter III, is based on the use of the laws and principles of classic mechanics, physics, and chemistry. Turbulence plays an important role. Contrary to the study of conventional fluid mechanics, the study of this system must also take into account phase changes in air and water, chemical reactions, and radiation effects. Thus a greater number of variables is introduced as well as a greater number of equations. Some simplifications become necessary for the models, or even the most powerful computers would be incapable of producing solutions. The behavior of the atmosphere and hydrosphere is, in fact, deterministic, even though the predictability of the system is limited as a result of difficulties in the practical calculations of its states, which require approximations, and imprecision in the observations of initial states. However, methods and techniques of measurement are steadily being improved (see chapter III).

Determinism may be discovered in other sciences, like astronomy, chemistry, and biochemistry, and also to varying degrees in certain specific problems in biology and, more rarely, in economics (Kline, 1987). In biology (chapter V), important recent discoveries show that determinism is inherent in genetic codes, while the laws of embryology illustrate with increasing cogency the marvelous organization of living creatures. In economics (chapter VI) mathematical models succeed only partially in representing quantifiable variables. Statistical methods are used, but these models cannot mathematize human choices.

In scientific research, one must employ a causalist or, when appropriate, a deterministic approach and follow it as far as possible. This methodology has enjoyed its greatest success in the area of physicomathematical determinism. It permits a complete local analysis of a system and, by analytical or numerical integration of the underlying equations, a reconstruction that yields all the physical properties of the entire system (see II.2). But to study the complex phenomena of many natural systems, such as physics, biology, and the social sciences, the purely causalist and deterministic method is insufficient to explain all the scientifically observed phenomena. Local analysis is not sufficiently complete to permit an integration of equations that would yield all the properties of the system (Kline, 1987). Thus complementary techniques of investigation and explanation — using *local* analysis and considering *global* properties for the whole system, from the initial conditions to the final state — have to be employed.

VII.3. The Interpretation of Phenomena according to Causality, Determinism, and Teleonomy through Regularization, Optimization, and Convergence toward Final States

The ancient idea that the whole is more than the sum of its parts can now be confirmed and given precise meaning. In a system, the change undergone by each element depends upon those undergone by the other elements and by the total system, which depends upon each of its constituents. This is illustrated by interactions between local and global properties, which form loops, or feedbacks. When a system being studied admits a mathematical model, the nonlinear terms reveal the existence of such interactions.

Efficient causes express the causal relations produced by local actions, whose consequences will affect the constituents of the system at some future time. As has been well known since Einstein, an instantaneous force cannot act from afar, because no interaction has a velocity greater than the speed of light. However, at a first approximation, one may consider that the gravitational forces in classic mechanics act instantaneously (see I.5).

A phenomenon that both precedes and determines another is an efficient cause, and following Kant, an entirety that produces the existence of its own parts is the true definition of teleonomy. Thus in embryology the egg is a composite of potentialities (see chapter V).

Teleonomy is characterized not only by a global tendency to change in a direction leading to a convergence toward a well-defined final state (or family of final states or final statistically permanent states) but also by the fact that this objective remains constant despite variations, within certain limits, of local efficient causes along the paths they follow, and this occurs subject to the rules of the local and global optimization of possible states (see I.5). As Julian Hunt (1990, 389) put it in his review of the French version of this book, "a nice point is made that the 'cause' of a final state of a complex system is a *convergence* of the individual causes. In other words, the final state in a physical system is in a *broad sense deterministic.*"

The representation of the changes of a dissipative dynamic system in a phase-space that includes attractor basins illustrates these teleonomic properties. Points situated within an attractor basin but representing different initial conditions have corresponding different trajectories, which nevertheless eventually tend toward the attractor—that is to say, toward a *final state.* Notice that the final state is a statistically permanent state, in the case of the atmospheric-hydrospheric system. This ultimate convergence is even recognizable for chaotic, dissipative systems, where trajectories that are initially very close and begin diverging rapidly eventually still tend toward the attractor, because dissipation contracts the volume in phase-space. In addition, turbulence strongly increases the dissipation and, thus, the convergence (see I.3 and I.5). To cite Hunt once more, "dissipative systems are much more weakly correlated with their initial conditions than nondissipative systems" (ibid., 387).

Within a system, relations between local properties and efficient causes, between global properties and teleonomy, and between elements and the entire system lead to reciprocal relations among the efficient causes and teleonomic behavior, which are thus complementary rather than antinomic.

In the case of systems that can be represented mathematically, and particularly physical systems, the causal character of their behavior can be expressed in differential equations, and their optimal and teleonomic qualities appear in variational formulations (see chapter IV). Physical phenomena obey the *principle of least action*, which, for a given cause, asserts that a particular criterion function must achieve its minimum—that is, the effects are *optimal.* For example, a body that is not subject to any force will move in a straight line, passing from one point to another by the shortest possible distance. Light and other radiations that are propelled from one point to another choose, as it were, their trajectory in such a fashion as to accomplish the trip in a minimum of time.

When solar radiation raises the temperature of Earth, the air in contact with the surface is warmed by molecular conduction; beyond that, if the differences in temperature exceed a certain instability threshold, then the air is set into motion and transfers greater quantities of heat by convection. In fact, as it grows warmer and expands, its density decreases. It becomes lighter than the surrounding air, rises to higher altitudes, where it is cooled, and then descends, permitting the cycle to recommence. Vortices form about the center of an ascending current and about the periphery of a descending one. They roll about each other with a minimum of friction. Great quantities of thermal energy — both sensible heat and the latent heat of water vapor — are thus transported with minimal dissipation of kinetic energy. The ultimate average effect of this procedure is the regulation of temperature upon the Earth's surface, in the atmosphere, and in the hydrosphere. Observations agree with the causalist and teleonomic physical theories that require an optimization of *possible* states.

More generally, taking into account turbulence and the phase changes of water and various chemical reactions, the behavior of the atmosphere and hydrosphere is governed by determinism and has average effects that regularize climate (see chapter III) Again, this is a teleonomic optimization process operating upon possible and compatible states. Without operating perfectly — because in fact these effects do not obviate certain larger local variations limited in time and called natural catastrophes — they maintain upon Earth, globally and within narrow limits, the conditions necessary for the preservation of life — for its development and evolution.

In biology, the genetic code and the laws of embryology ensure the development and regulation of living creatures and permit their evolution into increasingly complex forms (see chapter V). One could also call this an optimization of possible compatible states. This organization of living matter directs life globally in some teleonomic mode. The same observations regarding regularization mechanisms may be made in other domains, most notably in medicine and ecology. Thus the causalist, regularizing, optimizing, and teleonomic tendencies we have described in the behavior of nonliving and living matter are recognizable on a still higher level by the convergence of the physicochemical states of the terrestrial medium toward the many strict conditions necessary for the maintenance and development of life.

This compatibility cannot be attributed to chance, because to do so would imply a coincidence of so many independent series that the composite probability would be practically zero. Nor can it be more than only partially explained by the adaptation of animate creatures to their environment, because that is slow and is limited by various thresholds.

In conclusion, the behavior of the natural, nonliving milieu of the atmospheric-hydrospheric system and the behavior of living organisms depend upon causality and exhibit a convergence, through teleonomic coordination, with a bias in favor of life. This conclusion agrees with the ideas of Gottfried Leibniz (1968) that the series of efficient causes and the series of objectives agree completely among themselves as though the former were directed by the influence of the latter. The meaning of Leibniz's thought, *the best of all possible worlds*, reveals its great profundity. In theoretical physics (chapter IV), we have seen that Leibniz, who discovered differential calculus at the same time as Newton, also formulated a precursor of the calculus of variations. He thought about an optimization of possible and compatible states of the diverse parts of the world. He was not an optimist in the vulgar sense of the word but, rather, the first theoretician of optimization.

The problem is thus to know what criteria might yield a definition of that idea of *the best*. In a general and cosmic context, that question remains open. But from the human perspective, it seems natural to believe that the best of all possible worlds can only be one where life exists, flourishes, and perfects itself.

VIII

A Philosopher's Reflections

Jean Guitton

"Nature has exposed all its truths, each one apart. Our art has included them, the one in the others. This is not natural. Each has its own place." This observation by Blaise Pascal (1961, 183) will help me set out several reflections on this volume, where the authors have expressed experiences drawn from several disciplines in the hope of illustrating what Pascal calls "the always admirable relation that Nature has created between apparently disparate things."

According to Aristotle, the metaphysician is not satisfied merely to think; he "thinks his thoughts," he seeks to unify the diverse forms of being. Now, two ways exist of "thinking one's thoughts." The first considers human thoughts in their historical succession. The second seeks, beyond variations and contradictions, to discern a common, invariant, fundamental element in thought that remains identical despite differences among circumstances and mentalities where it is encountered.

None can deny that the problem posed for the intelligence by the ideas of causality and teleonomy can be found among all thinkers since the beginnings of philosophy in Greece. But the progress of physical science since the seventeenth century has lent it a sudden pertinence. In our time, at the end of the second millenium of the Christian era, these problems of causes and ends are more urgent than ever, even if they are less visible. Let us say that, for a mind trained in the mathematical, physical, biological, or social sciences, the category of *teleology* seems to be mythical and outmoded to the point that one must use prudish and even embarrassed words to reintroduce it where the weight of experiments suggests it.

In biology, language has been renovated. Ideas that seemed to support the ancient idea of *purpose* have largely been eliminated. Such expressions as *genetic information*, *coded program*, *model*, *gene*, *DNA*, and so on have appeared. And, by a complaisance that Jacques Monod (1970, 22) rightly describes as "an objective favor," a new word, *teleonomy*, has replaced the out-of-fashion *teleology*.

In this pluridisciplinary work, the authors have liberated themselves from that scruple and have dared to speak about teleonomy, which suggests to them the nonpurposive part of teleology, that nonmetaphysical part warranted by their scientific experience (see "Preface to the English Edition").

Taking the long view of the history of thought from the pre-Socratics to Heidegger and Michel Foucault, one can take either of two opposing directions. In the nineteenth century, Félix Ravaisson-Mollien (1884, 295) proposed this dichotomy. One of these directions runs from the bottom to the top. The other situates itself at the summit and descends from that height to plumb the depths.

The first hypothesis uses what Aristotle calls a *material cause*, while the second uses what Aristotle calls a *final cause*. In the first case, one does not speak about *material* in the vulgar sense of some amorphous mass but as a more abstract element of reality, one that, because it lacks form, may assume any form it may be lent. Aristotle enumerates the various aspects of this method, which defines the dialectic that is as immanent in the thought of Democritus, Epicurus, and Lucretius as it is in that of Darwin and Monod.

One begins with a multiplicity of autonomous, independent, discontinuous elements. One expresses the indivisibility and plurality of these elements by a Greek word, ἄτομος (*a-tom*, uncut). One studies their possible combinations according to the laws of random combinations and large numbers, all the while setting aside, as though in parentheses, any idea of direction, of convergence, of design. The element that Aristotle calls ὕλη (*hule*, material) is sufficient to explain the emergence of increasingly complex forms. Thus, higher forms are reduced to lower forms by mechanical causality. Of the innumerable possible combinations of atoms, those few that subsist are those adapted to their environment. Thus, to use modern language, *chance* and *necessity* are the masters of the game. Lucretius' poem explains this magnificently.

The second, metaphysical, hypothesis does not deny the existence of material causes, but it claims that such causes are inferior, directed, completed, and even sublimated by a superior cause called a *formal* or *final* cause. Gottfried Leibniz (1968) found a profound expression for that material within form, believing that inferior things exist within superior things in a nobler fashion than they can exist within themselves. One could cite the examples of mineral substances existing within vegetal substances, vegetal substances within animal substances, and animal substances within reasoning beings.

This was the consistent teaching of the metaphysicians—of Plato, Aristotle, and Plotinus; of the Christian philosophers; of the Cartesians; of Leibniz, Berkeley, and the Kantians; and of the modern philosophers, our contemporaries, for example, Henri-Louis Bergson and Pierre Teilhard de Chardin. In quite different terms, realists and idealists have held that the highest form of being is the first cause of beings, whether it be by emanation or creation. There is an echo of ψυχή, of mind, within all beings, even within the most inferior forms of organized material.

"Invisible harmony prevails over visible harmony," as Heraclitus observes, and the authors of this study, independently and by autonomous and even opposing methods, tend to the same conclusions. This tendency leads me to think that, in the current state of science, it is the second direction we have sketched that conforms most closely to experience.

We have not arrived at this result a priori or by bias. We have restricted ourselves to the scientific method, which has been sovereign in Europe for three centuries and which takes mechanism as a starting point and does not envisage a teleonomic solution until obliged to do so by experience. An a priori negation of all theory is the surest and most rigorous procedure. This is what Aristotle thought when he began his *Metaphysics* by criticizing all previous metaphysics, which denied what he was about to try to prove. It is clear that, by pride of place at the center of worlds, Occidental philosophy has indeed abused the principle of teleology and that that wrongheadedness has retarded the progress of science. Knowledge of the cosmos has been acquired in opposition to the anthropocentric teleology propagated by Christianity.

Aristotle conceived of his system while denying Democritus' atomism. It is doubtless by criticizing Monod's idea that biology will be renewed, since it is well known that progress is made by criticizing criticism.

Contemporary thought concentrates on negative concepts, nonbeing, disorder, privation, anxiety. If one wanted to seek an axiom common to these studies, one might formulate it as "Nonbeing exists, disorder exists." And several of our essays here indeed deal with chaos and disorder. More precisely, they deal with turbulence. It is from a point of departure in turbulence that we consider the existence or nonexistence of disorder.

Disorder is never anything but a different order than we expect. There exist levels in the nature of things, and phenomena change—not their natures but their appearance—according to the level, whether higher or lower, at which they are studied. In this sense, turbulence is imagined at either a microscopic or macroscopic level. So it was with turbulence as our guide that we accomplished this interdisciplinary research into the different stages of experience.

Let us take the concept of *chance* as an example. It is necessary here to distinguish what is anthropocentric in this rather muddled concept from that which expresses a reality inherent in the nature of things. Those who have studied chance from this point of view agree that there indeed was an objective chance each time the elements of the cosmos disassociated. Let us consider games of chance, which suggest probability theory; they suppose the existence of a microcosm whose elements and instants are disassociated, like Democritus' atoms. This random, discontinuous, and diverse universe—composed of a multitude of constituents, with each combination unrelated to the preceding one—is an image of what the universe is for these learned probabilists.

Does this mean that all order is banished from this universe? Having believed that an atomistic universe is in disorder and that nothing in it depends upon any laws other than those of large numbers, I now conclude that, in such a universe, there exists a superorder distinct from disorder. In effect, if I consider the world that Democritus imagined, where atoms fall into the void, collide, and form masses, I am struck by the fact that favored aggregates of masses subsist. This implies that nature has an obscure preference for those masses we call *bodies.* An aggregate formed by atoms at a given moment should disappear in the next instant, while ephemeral aggregates recur. Furthermore, they become increasingly perfect, from one moment to another. In fact, according to Lawrence Henderson (1921; see I.8, III.1.4, and VII.2), such coincidences would be much too numerous for us to see the effects of chance. The properties of elements exhibit a teleonomic arrangement.

Molecular motion, even when in apparent disorder, obeys the deterministic principles of classic mechanics (see I.5.7 and I.7.2). Many combinations have been sketched by the remarkable fecundity of nature, although only one of them is capable of subsisting. At the same time that the human species was emerging, there were multiple, random, aberrant variations among anthropoids. The human form appeared like the winning number in a lottery: once the lottery player sees the winning number turn up, he leaves the casino jubilant—he has won. This is the way nature works, once it has achieved, after turbulence and random variations, a form, however improbable, that permits further adaptation. What is inexplicable from the point of view of pure science—by which I mean science that does not admit teleonomy—is that evolution preserves some variations with a view to the future, to ensure that other, equally improbable variations will occur to complete it.

In the case of the origins of humans, evolution has preserved the first approximations to humanity despite their awkwardness. It is that continuity of random variations that favors a single line of descent—the possibility of repairing evolutionary errors, of reconstructing a living creature from a malformation (see V.1 and V.2)—that suggests the presence of teleonomy.

If we consider the work that produces poetic creation, as Paul Valéry, for example, explains it, we realize that a poet begins with disordered combinations, a turbulence of dreams (an apparent disorder), which he accepts and even pursues. And yet this disorder is governed by an alert mind, ready to seize a sonorous cadence, retain a maxim, or fix an admirable verse that sticks in the memory. Then the poet finds the form that seems to have preexisted because of its perfection but that could never have existed but for that primordial chaos. The priority of such "disorder" over order is favorable to progress, just as in games of chance the practice of shuffling the cards and cutting the deck before each draw introduces new combinations. When a piece of poetry has emerged from an antecedent chaos, most often it displays a greater beauty than one that is merely a poet's contrivance, and in the former case the poet himself is surprised by the poem's success.

One may compare it to a great passion. There, too, one sees that, among all the encounters of men and women that occur by accident, if a single one lasts and promises to endure it is because the bodies, or the minds, or their combination matched—and the relationship became permanent. This is why it is said, Love is a chance the heart has believed in.

Thus order and chaos appear to be complementary points of view. Independent phenomena may be disordered with respect to mechanics, which itself appears to be disordered when I glimpse a higher level of complexity—for example, in biology, where I recognize teleonomy. Nature is always ordered, but the densities of order it admits are variable. Such is also the case in the psychological universe. My dream is less ordered than my dreaming, my dreaming less than my reflection. Inductive reasoning is less ordered than deductive reasoning. The fantasy world of dreams, as Freud recognized, has a logic. The world of sin, Christian theologians tell us, has its laws, too. And it is possible that what we call evil may be a boon, and even the best of boons, when we see it in the totality of the world.

I hold that there exist two types of order in nature: one we call *order* or *law* or *necessity*, and it seems to us to require the exclusion, the casting into the abyss, into primordial chaos, of any other concatenation of nature's constituents; the other order consists of a crossing of series, of collisions of minimal causes, of quantum incertitude, and at a human scale, of encounters. There are two kinds of events, those that appear in independent series and those that result from the crossings of such series. And we see that nature, to the extent that it progresses, increases the number of random occasions, or rather it confers upon them a continually increasing significance. This is evident in the mechanism of biological conception, where randomness governs the fusion of the two elements that will constitute an embryo. There is in both nature and history a movement that, in the very heart of order, *imitates disorder* by liberating effect from cause, instant from instant, result from intention.

And since nothing stands in the way, in an expanding cosmos that includes knots, bifurcations, and crossovers, life and free will introduce their actions, and a calculation of what will occur will surpass the capacity of any finite intelligence. Now, since neither the idea one forms of physical order nor the experience one has of science is opposed to that lack of "determination," and because human objectives oblige us to suppose it, one may affirm, extrapolating from freedom of will to the conditions that make it possible, that *contingency* exists at the heart of being. The image that we now form of a science of nature that operates by laws on the coarse level of our experience—but that is, however, uncertain at an ultimate level of its elements—corresponds to what we have been able to deduce. There is no liberty if, within the substance of matter, there is no contingency.

The essays presented here have two tendencies, the one more mathematical and deterministic, the other more teleonomic. Our colleagues in physics and mathematics describe in rigorous language phenomena that are apparently indeterminate and teleonomic in order to reduce them to mathematical expressions. It is in this sense that turbulence is physically and mathematically determined and may be explained by a system of equations whose number equals that of the unknowns. But once this is said, it is no less true that when fluid flows are observed at different scales of measurement, when structures are transformed and regrouped, one recognizes new forms of order which are not reducible to disorder, because order may be complex and fluctuating and may change while changing scale. Thus, even on a purely physical and mathematical plane there exists a virtual teleonomy, which is all the more remarkable because it can be reduced to a calculation (see chapters I, II, IV, VII).

Of all the scientific disciplines, biology presents the most remarkable examples of teleonomy. Even if one could, in extremis, explain the general phenomena of reproduction by pure mechanics, this is not the case for the generation of organs.

Let us now return to the preceding considerations, seen in this new light. If the universe exhibits a regulatory tendency that permits life to emerge and perpetuate itself, if nature has a preference for the animate and the human, if we discover that same preference in living microcosms, down to the most elementary cells, is it not conceivable that teleonomy is present at virtually all levels? And consequently, are we not led to think that teleonomy will prevail over causality insofar as it is the global cause for the convergence of local causes? But this convergence of causes, this teleonomy that Henderson (1921) calls the fitness of the environment, can it be recognized by a scientist? Does it not require a different sort of intelligence than that of the pure scientist to conceive of it, to think about it, without causing scandal?

To employ the terms of cybernetics, it is a *feedback*, a *feedbefore*, a *cause in reverse*, a *retrocause*. Chance, necessity, and teleonomy are three solutions, and an elimination of the first leaves only the two others.

Chapter III leads us to suppose that there is such a teleonomy. It is turbulence that prevents the sun from incinerating Earth, because it interposes between it and us the turbulent atmospheric-hydrospheric system, which has permitted life to emerge upon this planet and to subsist here. Turbulence seems to be evil because it seems a lesser order, if not disorder; but to follow our theory of optimization, it is an evil leading to a greater good. It is not a question of partial teleonomy, limited to an organism or a single species — an axis of evolution — but rather a question of a global and total teleonomy. This cannot be considered by a scientist qua scientist, but it can interest someone who passes from physics to metaphysics, from science to metascience.

One question, a daunting question that is generally ignored, becomes urgent. To understand a phenomenon that is incompletely explained by causality, is it necessary to go beyond causes to an ultimate, unobservable cause? Do the achievements of science not oblige us to transcend science? One knows that organs regenerate when a segment retains completely or partially the potential that permits cells to differentiate and replace other cells. Now if the living organism can regenerate, it is because there exists in nature an improbable law that permits it to revive after the failure of a partial death. And thus it behaves like a regulated embryo with a missing member; the other members assist the missing member to regenerate itself. It is difficult to explain this phenomenon exclusively by external mechanical factors. Here, too, the scientist is obliged to have recourse to teleonomy.

In political economy, the difficulties encountered by both industrialized and developing countries relate to ideas of the same type. All these economies are subject to turbulence. Economic turbulence was first considered a temporary perturbation that deterred a return to the equilibrium upon which the science of economics was first based. The earliest theory of political economics took only causal relationships into account. With the persistence of turbulence, a new point of view has emerged. Thus the idea of teleonomy was introduced. One now wonders whether that idea, which once was excluded from scientific theorizing, ought not become a major preoccupation of the social sciences. Toward what objectives are we being led by a purely quantitative economic growth, which certain thinkers expect will provide solutions that, in fact, never appear? A new world is being born where knowledge and action will not be separated.

Until now, we have considered only special sectors. But the function of the philosopher is to examine the totality of being. The profound law that has guided us in this study, *that the totality existed before its components*, may be applicable here to its greatest extent. It makes the most exorbitant demand upon the mind, because it must avoid the slightest specificity in order to consider nothing but the totality. This is why the point of view of the philosopher is inherently different from those of scientists, whoever they may be. In other words, science is always some kind of physics, while thought is, above all, metaphysics.

At the moment where we are situated, we are led to consider what a scientist has called *the fitness of the environment.* Those who contemplate the starry heavens may have the impression that we are alone in an empty, silent universe that is indifferent to humans. Three hypotheses have been suggested to explain the improbable teleonomy of the environment: random combinations, necessity, and teleonomy. Even though it may be necessary for explanatory purposes, the convergence is not observable in isolation.

There is, however, a point of view above experience. Each scientist studies a set of facts where causality, necessity, and probability are visible. As a scientist, he or she does not see the convergence of these three independent lines of analysis. Much the same is true of painting: looking at a painting with the eye of a scientist, one sees only a chaos of impressive colors. It is the superior eye of the intelligence that recognizes the brush strokes, the harmony, and the beauty.

For science to be possible, there must be a certain simplicity in the elements the universe is composed of. If there were thousands of elements instead of only a hundred, chemistry would be impossible. If the laws of nature could not be summarized in several simple principles, if there were not a harmony of those necessary principles, science would not be possible. If evolution were perpetual mutation, biology would not be possible. By these examples—and we could add many more—it seems that the universe supposes a harmony favorable to the free and thinking being. We admit that may be, as Pascal calls it, a bet, but it is the only reasonable bet. The world, as Lachelier (1924) would say, is a thought that does not think suspended from a thought that does think.

It now occurs, as it has indeed occurred often in its long history, that philosophy, despairing of perceiving Being, is entering a phase where it is satisfied to recognize Appearances (phenomenology) or Existence (existentialism). It restricts itself to structure and, a limiting case, to language alone—which is to say, to signs (mathematics, logic, linguistics). We are living through an era of nominalism. A sign is not a sign unless it signifies something other than itself. This nominalism is a critical and purifying moment that foretells a new realism. Teleonomy is such a sign.

References

Académie des Sciences. 1980. *Les sciences mécaniques et l'avenir industriel de la France.* Paris: La Documentation Française.

Ancel, Paul, and Pierre Wintemberger. 1948. "Recherches sur le déterminisme de la symétrie bilatérale dans l'œuf des Amphibiens." *Suppl. Bull. Biol. de Fr. et de Belg.* Paris.

Aspect, Alain, and G. Roger Dalibard. 1982. *Physical Review* 49:1804–10.

Aubry, Nadine, Philip Holmes, John Lumley, and Emily Stone. 1988. "The Dynamics of Coherent Structures in the Wall Region of a Turbulent Boundary Layer." *Journal of Fluid Mechanics* 192:115–73.

Balasko, Yves. 1978. "Economic Equilibrium and Catastrophe Theory: An Introduction." *Econometrica* 46:557–69.

Bass, Jean. 1967. *Eléments de calcul des probabilités.* Paris: Masson.

Bénard, H. 1901. Les tourbillons cellulaires dans une nappe liquide transportant de la chaleur par convection en régime permanent. *Ann. Chimie Phys.*, 7th ser., 23:68.

Bergé, Pierre, Yves Pomeau, and Charles Vidal. 1984. *L'ordre dans le chaos. Vers une approche déterministe de la turbulence.* Paris: Hermann. Published in English as *Order within Chaos: Toward a Deterministic Approach to Turbulence.* Trans. Laurette Tuckerman. New York: John Wiley, 1984.

Bergson, Henri. 1954 (1907). "Evolution créatrice." In *Collection de Bibliothèque de philosophie.* 3d ed. Paris: Alcan.

Betchov, Robert, and William O. Criminale. 1967. *Stability of Parallel Flows.* New York: Academic Press.

Bordas, ed. 1983. *Dictionnaire de philosophie.* Paris.

Broglie, Louis de. 1937. *Matière et lumière.* Paris: Albin Michel.

Brun, Edmond, André Martinot-Lagarde, and Jean Mathieu. 1968. *Mécanique des fluides.* Paris: Dunod.

Bruter, Claude P. 1986. "Bifurcation and Continuity." In *Dynamical Systems: A Renewal of Dynamics*, ed. S. Diner, F. Fargue, and G. Lochak. Singapore: World Scientific.

Bureau des Longitudes. 1984. "La terre, les eaux, l'atmosphère." In *Encyclopédie scientifique de l'univers.* Paris: Gauthier-Villars.

Cartan, Elie. 1952 (1931). "Notice." In *Œuvres complètes.* Pt. I.v.I. Paris: Gauthier-Villars.

Chabreuil, Aline, and Marc Chabreuil. 1979. *Exploration de la terre par les satellites.* Paris: Hachette.

Chapman, Sydney, and T. G. Cowling. 1939. *The Mathematical Theory of Non-uniform Gases.* Cambridge: Cambridge University Press.

Darnell, John. 1983. "La maturation des ARN." *Pour la Science* 74:36–46. Published in English as "The Processing of RNA." *Scientific American* 249:72–82.

Delattre, Pierre. 1986. "An Approach to the Notion of Finality according to the Concept of Qualitative Dynamics." In *Dynamical Systems: A Renewal of Dynamics*, ed. S. Diner, F. Fargue, and G. Lochak. Singapore: World Scientific.

Dodé, Maurice. 1979. *Introduction à la mécanique statistique.* Paris: Hermann.

Emmons, H. W. 1951. "The laminar-turbulent transition in a boundary layer." *Journal of Aeronautical Science* 18:490.

Favre, Alexandre. 1946. *Appareil de mesures de la corrélation dans le temps.* Paris: Ministère de la Production Industrielle. Brevet Invention 6.B.12—CL3. No. 924800, Paris. U.S. Patent No. 2,693,908 (26 Aug. 1947).

———. 1958. "Equations statistiques des gaz turbulents." *Comptes Rendus de l'Académie des Sciences* séries A 246: 2576–79, 2723–25, 2839–42, 3216–19.

———. 1965a. "Equations des gaz turbulents compressibles." *Journal de mécanique* 4:361–421.

———. 1965b. "Review of Space-Time Correlations in Turbulent Fluids." *Journal of Applied Mechanics* series E32:261–90. Published in Russian in *Mekhanika* 2*90. Moscow: Editions MIR, 70–99.

———. 1969. "Equations statistiques des gaz turbulents." In *Problems of Hydrodynamics and Continuum Mechanics*, ed. USSR Academy of Science. Moscow: Nauka. Published in English as "Statistical Equations of Turbulent Gases." In *Problems of Hydrodynamics and Continuum Mechanics.* Philadelphia: Society for Industrial and Applied Mathematics.

———. 1983. "Turbulence: Space-Time Statistical Properties and Behaviour in Supersonic Flows." *Physics of Fluids* 26:2851–63.

———. 1992. "Formulation of the Statistical Equations of Turbulent Flows with Variable Density." In *Studies of Turbulence*, ed. Thomas Gatski, Sutano Sarkar, and Charles Speziale. New York: Springer.

Favre, Alexandre, Henri Guitton, Jean Guitton, André Lichnerowicz, and Etienne Wolff. 1988. *De la causalité à la finalité. A propos de la turbulence.* Paris: Maloine.

Favre, Alexandre, and Klauss Hasselmann, eds. 1978. Vol. I. *Turbulent Fluxes through the Sea Surface: Wave Dynamics and Prediction.* New York: Plenum.

Favre, Alexandre, Leslie S. G. Kovasznay, Régis Dumas, Jean Gaviglio, and Michel Coantic. 1976. *La turbulence en mécanique des fluides.* Paris: Gauthier-Villars.

Fourastié, Jean. 1979. *Les trente glorieuses—1946–1975.* Paris: Fayard.

Glansdorff, P., and Ilia Prigogine. 1971. *Structure, stabilité et fluctuation.* Paris: Masson.

Grünbaum, Adolf. 1973. *Philosophical Problems of Space and Time.* Boston: Reidel.

Guitton, Jean. 1953. *Pascal et Leibniz.* Paris: Aubier.

Hansen, J. P., and D. Lévesque. 1985. "La dynamique moléculaire." *Le Courrier du C.N.R.S.* supplement, 59.

Henderson, Lawrence J. 1921. "La finalité du milieu cosmique." *Bulletin de la Société Française de Philosophie* 16:1–29.

Hinze, J. O. 1975. *Turbulence.* New York: McGraw-Hill.

Howarth, Leslie, ed. 1956. Vol. I. *Modern Developments in Fluid Dynamics with High Speed Flow.* Oxford: Clarendon.

Hunt, Julian C. R. 1990. "De la causalité à la finalité. A propos de la turbulence." *European Journal of Mechanics* 9:385–92.

Hunt, Julian C. R., and David J. Carruthers. 1990. "Rapid Distortion Theory and the 'Problems' of Turbulence." *Journal of Fluid Mechanics* 212.

Jacob, François. 1976. *La logique du vivant.* Paris: Gallimard.

Jacob, François, and Jacques Monod. 1961. "On the Regulation of Gene Activity." *Cold Spring Harbor Symposium on Quantitative Biology* 26.

Karman, Theodore von, and Th. Howarth. 1938. "On the Statistical Theory of Isotropic Turbulence." *Proceedings of the Royal Society of London* 165:192.

Kline, Stephen J. 1987. "The Logical Necessity of Multidisciplinarity: A Consistent View of the World." *Bulletin of Science, Technology, and Society* 6:1–26.

Kolmogorov, Andrei N. 1941. "Local Structure of Turbulence in an Incompressible Fluid for Very High Reynolds' Numbers." *Doklady Akademia Nauk SSSR* 30: 299–303.

———. 1950. *Foundations of Probability.* New York: Chelsea.

———. 1962a. "Précisions sur la structure locale de la turbulence dans un fluide visqueux aux nombres de Reynolds élevés." *Colloque International CNRS*, no. 108, 452–58.

———. 1962b. "A Refinement of Previous Hypotheses Concerning the Local Structure of Turbulence. . . ." *Journal of Fluid Mechanics* 13:82–85.

Lachelier, Jules. 1924. *Du fondement de l'induction.* Paris: Alcan.

Lacombe, Henri. 1965. *Cours d'océanographie physique.* Paris: Gauthier-Villars.

Lalande, André. 1983. *Vocabulaire technique et critique de la philosophie.* Paris: Presses Universitaires de France.

Landis, Fred, and Ascher Schapiro. In Van Dyke, 1982.

Laplace, Pierre-Simon de. 1921 (1814). "Essai philosophique sur les probabilités." In *Edition des maîtres de la pensée scientifique.* Paris: Gauthier-Villars.

Larousse. 1979. *Petit Larousse illustré dictionnaire encyclopédique.* Paris: Larousse.

Leibniz, Gottfried Wilhelm von. 1963. *Basic Writings.* La Lalle, Ill.: Open Court.

Leontief, Wassily. 1968. *Essays in Economics.* Oxford: Oxford University Press.

———. 1974. *Essais d'économique.* Paris: Calmann-Lévy.

Leray, Jean. 1934. "Sur le mouvement d'un liquide visqueux remplissant l'espace," *Acta Mathematica* 63:193–218.

Libby, Paul A., and Forman A. Williams, eds. 1980. *Turbulent Reacting Flows.* Topics in Applied Physics 44. New York: Springer.

Lichnerowicz, André, François Perroux, and Gilbert Gadorff. 1976. *Structure et dynamique des systèmes. Séminaires interdisciplinaires du Collège de France.* Paris: Maloine-Doin.

Lorenz, Edward N. 1963. "Deterministic Nonperiodic Flow." *Journal of Atmospheric Sciences* 20:130–41.

——. 1969a. "The Predictability of a Flow which Possesses Many Scales of 'Motion.' " *Tellus* 21:289–307.

——. 1969b. Ed. *Proceedings of the Congress on Predictability and Fluid Motion.* Berkeley: University of California Press.

Lovelock, J. E. 1979. *Gaia: A New Look at Life on Earth.* Oxford: Oxford University Press.

Lumley, J. L. 1978. "Computational Modelling of Turbulent Flows." *Advances in Applied Mechanics.*

Lutz, Hubert. 1949. "Sur la production expérimentale de la polyembryonie et de la monstruosité double chez les oiseaux." *Arch. Anat. Micros. et Morph. expér.* 38:79–144.

Margenau, Henry. 1978. *Physics and Philosophy: Selected Essays.* Boston: Reidel.

Mathieu, Jean-Paul, Alfred Kastler, and Pierre Fleury. 1983. *Dictionnaire de physique.* Paris: Masson et Eyrolles.

Mayr, Ernst. 1961. "Cause and Effect in Biology." *Science* 134:1501–6.

——. 1974. "Teleological and Teleonomic: A New Analysis." In *Methodological and Historical Essays in the Natural and Social Sciences.* Vol. 14, *Boston Studies in the History of Sciences,* ed. Robert Cohen and Max W. Wartofsky. Boston: Reidel.

Medawar, Peter B., and Jean S. Medawar. 1977. *The Life Science: Current Ideas of Biology.* New York: Harper Row.

Mieghem, J. van., and L. Dufour. 1948. *Thermodynamique de l'atmosphère.* Mémoires 30. Brussels: Institut Royal Météorologique de Belgique.

Monod, Jacques. 1970. *Le hasard et la nécessité.* Paris: Seuil.

Monod-Herzen, Gabriel. 1976. *L'analyse dimensionnelle et l'épistémologie. Recherches interdisciplinaires.* Paris: Maloin-Doin.

Moreau, René. 1990. "Magnetohydrodynamics." In *Fluid Mechanics and Its Applications.* Vol. 3. Dordrecht: Kiwer.

Noether, Emmy. 1971 (1918). "Invariant Variation Problems." Trans. M. A. Tavel. *Transport Theory and Statistical Physics* 1:186–207.

O'Grady, Richard, and Daniel Brooks. 1985. "Teleology and Biology." In *Entropy, Information, and Evolution: New Perspectives on Physical and Biological Evolution,* ed. Bruce Weber, David Depew, James Smith, and Bradford Book. Cambridge: MIT Press.

Onion, C. T., ed. 1944. *The Shorter Oxford English Dictionary.* 3d ed. Oxford: Clarendon Press.

Pareto, Vilfredo. 1896. *Cours d'économie politique.* Paris: Sorbonne.

Parrot, J. L. 1985. *La fin et les moyens.* Paris: Maloine.

Pascal, Blaise. 1961 (1888). *Pensées et opuscules,* ed. Léon Brunschvicg. No. 21. Paris: Hachette.

Phillips, Owen M. 1977. *The Dynamics of the Upper Ocean.* Cambridge: Cambridge University Press.

Pittendrigh, C. S. 1958. "Adaptation, Natural Selection, and Behavior." In *Behavior and Evolution,* ed. A. Roe and G. Simpson. New Haven: Yale University Press.

Poincaré, Henri. 1908. *Science and Method.* London: Nelson.

Prigogine, Ilia. 1980. *From Being to Becoming.* Brussels: Freeman.

———. 1982. *Physique, temps, et devenir.* Paris: Masson.

Prigogine, Ilia, and Isabelle Stengers. 1979. *La nouvelle Alliance.* Paris: Gallimard.

Ravaisson-Mollien, Félix. 1884. "Rapport sur le prix Victor Cousin." In *La philosophie en France au XIXe.* 2d ed. Paris: Hachette.

Rebuffet, Pierre. 1968. *Aérodynamique expérimentale.* Paris: Dunod.

Royer, Daniel, and Gilbert Ritschard. 1985. "Portée et limites de méthodes qualitatives d'analyse structurale." *Revue d'Economie Politique* 95:777–94.

Ruelle, David. 1978. "Dynamical Systems with Turbulent Behavior." *Mathematical Problems in Theoretical Physics: Lecture Notes in Physics* 80:341–60.

Ruelle, David, and Floris Takens. 1971. "On the Nature of Turbulence." *Journal of Mathematical Physics* 20:167–92 and 23:343–44.

Saint-Hilaire, Etienne Geoffroy. 1822. *Philosophie anatomique. Des monstruosités humaines.* Vol. 1. Paris.

Schlichting, Hermann. 1955. *Boundary Layer Theory.* New York: McGraw-Hill.

Spemann, Hans. 1901, 1902, 1903. "Entwicklungsphysiologische Studien am Tritonei." *Arch. f. entw. Mech.* 12:224–65; 15:448-535; 16:551–632.

Swinney, Harry L., and Jerry P. Gollub, eds. 1981. "Hydrodynamic Instabilities and the Transition to Turbulence." In *Topics in Applied Physics.* New York: Springer.

Taylor, Geoffrey I. 1921. "Diffusion by Continuous Movement." *Proceedings of the London Mathematics Society* A.20, 196.

———. 1935. "Statistical Theory of Turbulence." *Proceedings of the Royal Society of London* 151:429–54.

Thom, René. 1980. "Halte au hasard. Silence au bruit." *Le débat* 4:119–33.

Townsend, A. A. 1976. *Structure of Turbulent Flow.* Cambridge: Cambridge University Press.

Truesdell, Clifford Ambrose. 1974. *Introduction à la mécanique rationnelle des milieux continus.* Paris: Masson.

Van Dyke, Milton. 1982. *An Album of Fluid Motion.* Stanford: Stanford University Press.

Watson, James, and Francis Crick. 1953. *Cold Spring Harbor Symposium on Quantitative Biology* 18:123.

Wolff, Etienne. 1933. "Recherches sur la structure d'Omphalocéphales obtenus expérimentalement." *Arch. Anat., d'Hist., et d'Embryol.* 16:135–93.

———. 1936. "Les bases de la tératogenèse expérimentale des Vertébrés amniotes d'après les résultats de méthodes directes." *Arch. Anat., d'Hist., et d'Embryol.* 22:1–382.

———. 1983. "Actualité d'un problème. la Finalité." In *Les sciences et la vie*, ed. E. Barreau. Paris: CNRS.

———. 1984. "La finalité en embryologie." *Revue des deux mondes* (August): 269–79.

Index

RELATED WORKS BY THE AUTHORS

Alexandre Favre, of the Academy of Science

"Equations des gaz turbulents compressibles." *Journal de mécanique* (Paris) 4 (1965): 361–421.

"Statistical Equations of Turbulent Gases." In *Problems of Hydrodynamics and Continuum Mechanics.* Philadelphia: Society for Industrial and Applied Mathematics, 1969.

La turbulence en mécanique des fluides, with L. S. G. Kovasznay, R. Dumas, J. Gaviglio, and M. Coantic. Paris: Gauthier-Villars, 1976.

"Turbulence: Space-Time Statistical Properties and Behaviour in Supersonic Flows." *Physics of Fluids* 26 (1983): 2851–63.

Henri Guitton, of the Academy of Moral and Political Sciences

"De l'imperfection en économie." In *Perspectives de l'économique, critique.* Paris: Calmann-Lévy, 1979.

Le sens de la durée. Paris: Calmann-Lévy, 1985.

Jean Guitton, of the French Academy and the Academy of Moral and Political Sciences

L'absurde et le mystère. Paris: Desclée de Brouwer, 1986.

Philosophie. Vol. 4 of *Œuvres complètes. Bibliothèque Européenne.* Paris: Desclée de Brouwer, 1978.

André Lichnerowicz, of the Academy of Science

Eléments de calcul tensoriel. Paris: Armand Colin, 1947.

Relativistic Hydrodynamics and Magnetohydrodynamics. New York: Benjamin, 1968.

Théorie globale des connexions. Rome: Edizioni Cremonese, 1955.

Théories relativistes de la gravitation et de l'électromagnétisme. Paris: Masson, 1954.

Etienne Wolff, of the French Academy and the Academy of Science

Les chemins de la vie. Paris: Hermann, 1969.

"La culture d'organes embryonnaires *in vitro.*" *Revue scientifique* 90 (1952): 189–98.

La science des monstres. Paris: Gallimard, 1948.